Conference of Theoretical Physics and Nonlinear Phenomena 2017

Journal of Physics: Conference Series Volume 988

Jakarta, Indonesia
1 August 2017

ISBN: 978-1-5108-6069-8
ISSN: 1742-6588

Printed from e-media with permission by:

Curran Associates, Inc.
57 Morehouse Lane
Red Hook, NY 12571

Some format issues inherent in the e-media version may also appear in this print version.

Printed by Curran Associates, Inc. (2018)

Institute of Physics
Dirac House, Temple Back
Bristol BS1 6BE UK

Phone: 44 1 17 929 7481
Fax: 44 1 17 920 0979

techtracking@iop.org

Additional copies of this publication are available from:

Curran Associates, Inc.
57 Morehouse Lane
Red Hook, NY 12571 USA
Phone: 845-758-0400
Fax: 845-758-2633
Email: curran@proceedings.com
Web: www.proceedings.com

Table of contents

Volume 988

Conference on Theoretical Physics and Nonlinear Phenomena 2017

1 August 2017, Jakarta, Indonesia

Accepted papers received: 14 March 2018
Published online: 28 March 2018

Preface

Papers

IOP Conf. Series: Journal of Physics: Conf. Series **988** (2018) 011001 doi:10.1088/1742-6596/988/1/011001

Preface

The Conference on Theoretical Physics and Nonlinear Phenomena is an annual meeting on theoretical physics and nonlinear phenomena in Indonesia. The inaugural of this conference was held in 2004, with the name of Workshop on Theoretical Physics. This year, the conference took place in the campus of Parahyangan Catholic University, Bandung, West Java. It is the first time for this campus to host this conference, where theoretical physicists from various places of Indonesia gathered and reported their researches progress. The topics presented in this conference covered theoretical physics in a broad range, including nonlinear science. Even more, in the last couple of the series, there were also talks reporting the development of experimental high energy particle physics, which were highly appreciated as a guide for theorists to make contacts with the real world.

There were 3 invited talks and 21 contributed talks presented in the conference. Thirteen years after its launch, the original spirit of the conference is unchanged, which is to gather Indonesian theoretical physicists to share their works and excitements. In particular, this conference can be also the place for the younger generations of theorists for presenting their ideas and grasping knowledges from the senior researchers.

We would like to deeply acknowledge all the parties involved in making this conference successfully held: the session chairs, the UNPAR physics students who were influential in assisting many aspects of the conference, and Parahyangan Catholic University who acted as the host institution for this conference. The complete contributions of the conference will be available as IOP Conference Proceedings.

Finally, we dedicate this proceeding to the late Prof. Dr. Asan Damanik who passed away a few days after the conference. His contributions to the Indonesian physics community is invaluable. We hope that his enthusiasm becomes an example for the younger generation of Indonesian theoretical physicists.

Co-chairs of the Conference
Haryanto M. Siahaan
Reinard Primulando

IOP Conf. Series: Journal of Physics: Conf. Series **988** (2018) 011002 doi:10.1088/1742-6596/988/1/011002

Peer review statement

All papers published in this volume of *Journal of Physics: Conference Series* have been peer reviewed through processes administered by the proceedings Editors. Reviews were conducted by expert referees to the professional and scientific standards expected of a proceedings journal published by IOP Publishing.

IOP Conf. Series: Journal of Physics: Conf. Series **988** (2018) 012001 doi:10.1088/1742-6596/988/1/012001

Search for the non-canonical Ising spin glass on rewired square lattices

Tasrief Surungan

E-mail: tasrief@unhas.ac.id
[1]Department of Physics, Hasanuddin University, Makassar 90245, Indonesia

Abstract. A spin glass (SG) of non-canonical type is a purely antiferromagnetic (AF) system, exemplified by the AF Ising model on a scale free network (SFN), studied by Bartolozzi et al. [Phys. Rev. B73, 224419 (2006)]. Frustration in this new type of SG is rendered by topological factor and its randomness is caused by random connectivity. As an SFN corresponds to a large dimensional lattice, finding non-canonical SG in lattice with physical dimension is desireable. However, a regular lattice can not have random connectivity. In order to obtain lattices with random connection and preserving the notion of finite dimension, we costructed rewired lattices. We added some extra bonds randomly connecting each site of a regular lattice to its next-nearest neighbors. Very recently, Surungan *et al.,* studied AF Heisenberg system on rewired square lattice and found no SG behavior [AIP Conf. Proc. 1719, 030006 (2016)]. Due to the importance of discrete symmetry for phase transition, here we study similar structure for the Ising model (Z_2 symmetry). We used Monte Carlo simulation with Replica Exchange algorithm. Two types of structures were studied, firstly, the rewired square lattices with one extra bonds added to each site, and secondly, two bonds added to each site. We calculated the Edwards-Anderson paremeter, the commonly used parameter in searching for SG phase. The non-canonical SG is clearly observed in the rewired square lattice with two extra bonds added.

1. Introduction

Spin glass (SG) system is an interesting subject to study, both theoretically and experimentally due to its wide range applications. It is considered for example as a candidate material for giant magneto-resistance (GMR)[1]. From the view point of computational (theoretical) physics, SG is a very good ground test for several such complex problems as NP-hard and optimization [2, 3]. This type of system is different from the usual magnet as it does not possess total magnetization at any temperature. It is grouped as a magnetic system due to the frozen random spin configuration at low temperatures. The frozen configuration is regarded as a typical ordered, called a temporally ordered phase rather than spatially ordered phase [4]. The SG phenomenon was first reported in experiment in the early 70s by Cannella and Mydosh who observed the presence of cusp instead of sharp peak in ac-susceptibility [5] of alloys composed from transition metal impurities hosted by nobel metals ($CuMn$ and $AuFe$). In these materials, magnetic impurities (Mn and Fe) with small concentration randomly occupy the sites of the non-magnetic hosts (Cu and Au).

Frustration and randomness are two main ingredients for a system to exhibit SG phase. Spins are frustrated if fail to simultaneously satisfy all their nearest neighbors in minimizing the interaction energy. This is firmly exemplified by spins on an AF triangular unit. Spins with ferromagnetic (FM) couplings (anti-ferromagnetically) will prefer to align in parallel to each other and anti-parallel fo AF couplings, in order to minimize the free energy.

The ground state (GS) configuration of the system is highly degenerate due to random frustration; leads the system to possess GS entropy, well known as residual entropy. It is also due to this factor that makes the system exhibiting many remarkable phenomena such as aging, rejuvenation and memory effects [6]. Due to the existence of a large number of local minima in the energy landscape, finding the GS energy of the systems is challenging for many optimization algorithms[7, 8]. Most SGs studied in literature are of canonical type, being characterized of having both FM and AF interactions. The so-called non-canonical SG has recently received significant interest [9, 10, 11, 14]. It is a new type of SG which is is purely AF, possessing random connectivity and its frustration is brought by the topological factor. This kind of SG was firstly reported for the Ising model on scale free network (SFN)[9]. Recently, we studied Heisenberg model on SFN with AF interaction and found SG phase transition[11].

The SFN corresponds to large dimensional regular lattices. A profound behavior of such network is small-world connectivity, where two spins randomly chosen are always connected by a shortcut distance in maximum, for example, of six units[13]. With this behavior, the notion of spatial dimension is lost. It is therefore desirable to probe this type of SG in reasonable spatial dimension. Very recently we studied Heisenberg model on rewired square lattice and found no SG transition[14]. Due to the importance of discreteness in phase transition [15, 16], here we study Ising model on rewired square lattices. The remaining part of the paper is organized as the following: We describe in Section II the model and method; discuss the result is in Section III. Section IV is devoted for summary and concluding remarks.

2. Model and Simulation

We study the Ising model which is written in the following Hamiltonian

$$H = -J \sum_{\langle ij \rangle} s_i s_j \tag{1}$$

where J characterizes the interaction between the Ising spin s_i residing respectively on i-th site of the lattice. We consider a pure AF system, so that $J < 0$. The summation is performed over all nearest neighbor spins, including the newly added neighbors due to rewiring. Regular lattices such as the square and cubic, all spins have the same number of neighbours, associated with a definite coordination number. Here, we studied two types of structures: (i) rewired square lattice with one extra bond added to each spin, and (ii) two extra bonds added. The first added bond connects a spin to one of its diagonal neighbours, while the second one connects each spin to one of the next-next nearest spins. Therefore, for the first structure, spins can have five, six or seven bonds. Particular realizations of these lattices are shown in Fig. 1.

The rewired lattices can be regarded as a quasi regular lattice. By excluding the double counting, the average number of bonds (ANOB) for the first and second structure is 3.0 and 4.0, respectively. In principle one can generate any lattice with fractional ANOB. As indicated in in Fig. 1, abundance of triangular units exist, rendering frustration due to AF couplings. Adding extra bonds up to certain number corresponds to increasing the degree of frustration. In principle, it is possible to define the degree of frustration and study its effect on SG properties. This is done for example in Ref. [10] which randomly rewired any pairs of spins. This procedure of rewiring totally breaks the translational symmetry of regular lattices. We do not take such procedure as the notion of spatial dimension needs to be preserved. Although the current developed model is purely theoretical, it may be implemented in growing AF-based SG material by introducing impurity, which was done for example in several References[19, 20, 21].

We performed Monte Carlo (MC) simulation using the Replica Exchange algorithm [17]. This algorithm is exceptionally suitable to overcome the slow dynamics due to the existence

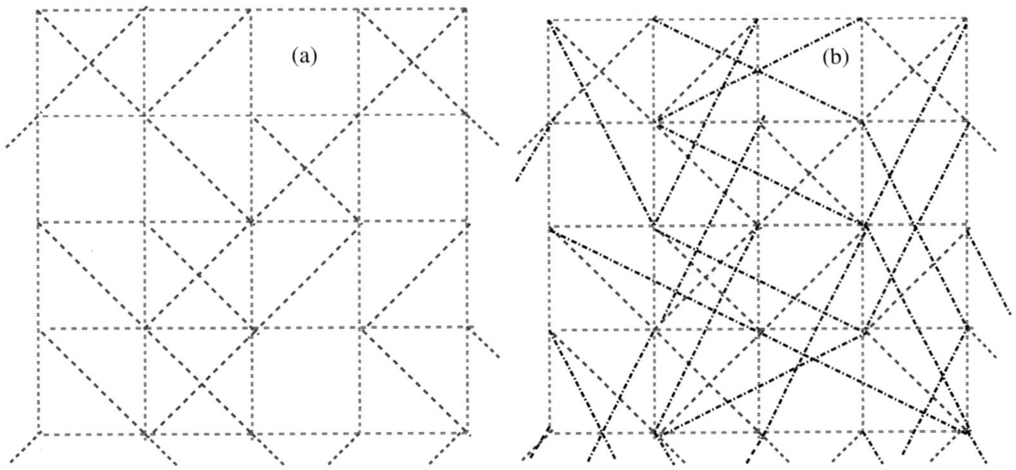

Figure 1. (Color online) Examples of rewired square lattices with (a) an extra bond and (b) two extra bonds; added to each spin. The bonds represented by dashed lines randomly connect each spin to the next-nearest or the next-next nearest neighbors.

of local minima in the energy landscape. The slow dynamics is a common problem in dealing with complex systems such as SGs where a random walker tends to get trapped at certain local minimum. The basic procedure of this algorithm is to duplicate the original systems into several replicas, each will be independently treated by a standard Metropolis algorithm. The final results of the calculation is obtained by combining all those separated calculations. During the calculation, replicas belonging to two adjacent temperatures are exchanged; enabling a random walker to stay away from any local minima.

If we start with M replicas; each is in equilibrium with a heat bath of temperature, then given a set of invers temperatures β, the distribution of probability in finding the entire system in a state $\{X\} = \{X_1, X_2, \ldots, X_M\}$ is given by,

$$P(\{X, \beta\}) = \prod_{m=1}^{M} \tilde{P}(X_m, \beta_m), \tag{2}$$

with

$$\tilde{P}(X_m, \beta_m) = Z(\beta_m)^{-1} \exp(-\beta_m H(X_m)), \tag{3}$$

and the partition function $Z(\beta_m)$ is associated with the m-*th* replica. An exchange matrix between replicas is defined as $W(X_m, \beta_m | X_n, \beta_n)$, which is the probability to exchange the configuration X_m of β_m with the configuration X_n of β_n. With the requirement of preserving the entire system at equilibrium, we apply the detailed balance condition to the transition matrix

$$\begin{aligned} &P(\{X_m, \beta_m\}, \ldots, \{X_n, \beta_n\}, \ldots) \cdot W(X_m, \beta_m | X_n, \beta_n) \\ &= P(\{X_n, \beta_m\}, \ldots, \{X_m, \beta_n\}, \ldots) \cdot W(X_n, \beta_m | X_m, \beta_n), \end{aligned} \tag{4}$$

along with Eq. (3), so that we have

$$\frac{W(X_m, \beta_m | X_n, \beta_n)}{W(X_n, \beta_m | X_m, \beta_n)} = \exp(-\Delta), \tag{5}$$

where $\Delta = (\beta_n - \beta_m)(H(X_m) - H(X_n))$. With this constraint, we can choose the matrix coefficients following the standard Metropolis algorithm which gives

$$W(X_m, \beta_m | X_n, \beta_n) = \begin{cases} 1 & \text{if} \quad \Delta < 0, \\ \exp(-\Delta) & \text{if} \quad \Delta > 0. \end{cases} \tag{6}$$

Since the ratio of acceptance exponentially decays with $(\beta_n - \beta_m)$, the exchanges of temperatures are allowed only between temperatures next to each other, e.g., the terms $W(X_m, \beta_m | X_{m+1}, \beta_{m+1})$.

In doing the simulation, one MC step (MCS) is defined as randomly visiting all spins at at each replica, either consecutively or randomly, and perfoming update based on MC rule. After a number of MCSs, pairs of configurations belonging to corresponding temperatures next to each other are exchanged following the probability written in Eq. 6. Two consecutive exchanges are assigned as odd and even. For an odd exchange, we swap the pairs of replicas X_1 and X_2; X_3 and X_4; $\cdots$; X_{M-1} and X_M; while for the even one we exchange X_2 and X_3; $\cdots X_{M-2}$ and X_{M-1}.

For each realization of connectivity we start from a random spin configuration. Then, we run 10^3 MCSs to equilibrate the system before taking a total of 4×10^4 sample points for the ensemble averages. Several extra MCSs between sample points were taken to avoid temporal correlation. We probe the high and the low temperatures to search for SG phase transition. The ensemble averages obtained for each size of lattice then averaged over many lattice connectivities, which is a standard procedure in doing MC simulation for a random system such as SG. In the next section, we present the results of our study.

3. Results and Discussion

3.1. Energy and the specific heat

We have simulated the AF Ising model on randomly rewired square lattices with ANOB = 3.0 and 4.0 for several linear sizes $L = 16, 24, 32, 48$, and 64. The periodic boundary condition is applied so that every site of the original square lattice possesses similar number of nearest neighbors. For each system size we took large number of realizations of connectivities. The result is obtained by averaging over these realizations. Each realization is associated with one particular randomly generated connectivity distribution. For the results to have better statistics, we took large number of realizations. Previous study of Heisenberg SG on SFN took 1000 realizations. Here we took smaller number of realizations (N=250) due to its degree of randomness is less compared to the previous system.

To probe the equilibrium properties which is essential for the calculation ensamble average of physical quantities, we evaluated the energy time series (ETS). The average energy and specific heat can be extracted from ETS. Whether the system is in equilibrium or not can be characterized by analysing the energy time series. In MC simulation, time is associated with series of MCS's. We run M_s MCS's for every temperature and take N sample points out of M_s. To guarantee that the system is well equilibrated, initial MCSs is performed before doing measurement, normally around 10000 MCSs.

From the energy time series (ETS), We extracted two quantities, namely the ensemble average of energy, $\langle E \rangle = \frac{1}{N}\sum_N E_i$, and the specific heat is written as the following

$$C_v = \frac{N}{kT^2}\left(\langle E^2 \rangle - \langle E \rangle^2\right) \tag{7}$$

where k and N are the Boltzmann constant and the number of spins, respectively. The temperature dependence of this quantity is shown in Fig. 2 and Fig. 3.

The temperature dependence of specific heat does not have diverging peak for lattice with ANOB = 3.0, and increasing with system size for lattice with ANOB = 4.0. This clearly signify

IOP Conf. Series: Journal of Physics: Conf. Series **988** (2018) 012001 doi:10.1088/1742-6596/988/1/012001

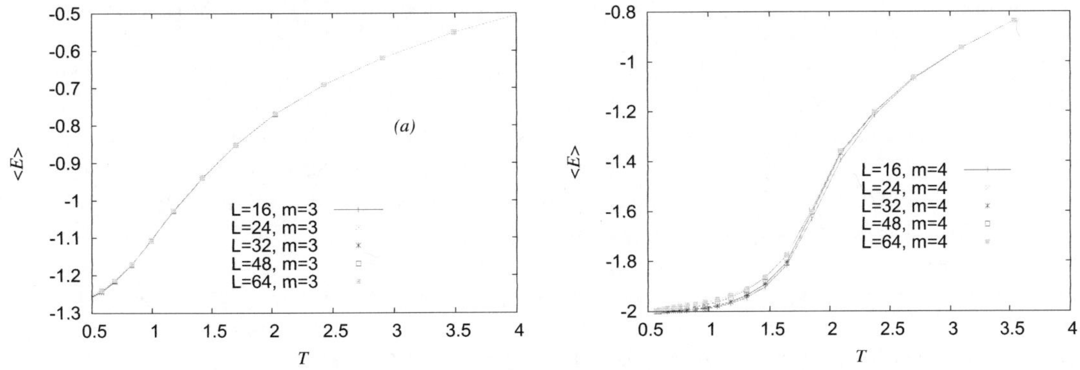

Figure 2. Temperature dependence of energy of lattices with (a) ANOB = 3.0 and (b) 4.0 for various system sizes.

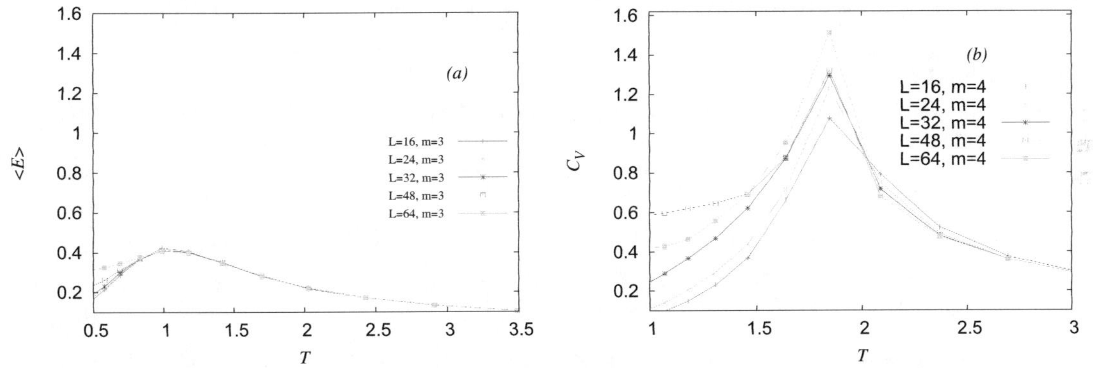

Figure 3. Temperature dependence of specific heat of lattices with (a) ANOB = 3.0 and (b) 4.0 for various system sizes.

an absence of (the presence of) phase transition in the first (second) type of lattice. For a more convincing analysis, we calculate the Edwards-Anderson parameter which is presented in the next sub-section.

3.2. Spin Glass Order Parameter

The existence of SG phase transition is characterized by the overlapping parameter, also known as Edwards-Anderson (EA) parameter [18] defined as follows

$$q_{EA} \equiv \left[\langle | \frac{1}{N} \sum_i s_i^{(\alpha)} s_i^{(\beta)} | \rangle \right]_{\mathrm{av}} , \tag{8}$$

where two sets of systems s_i^α and s_i^β having similar connectivity distribution. The aim of this parameter is to detect the frozen configuration when system is in SG phase; which is merely introduced for numerical simulation. It is almost impossible in experiment to obtain two duplicated systems with similar coupling distribution.

For Ising case, the EA parameter (q_{EA}) is much simpler compared to the Heisenberg case (vector spins) which involves tensor product [11, 14]. If system is frozen, EA parameter will be finite and vanish otherwise. The temperature dependence of q_{EA} for various sizes, $L = 16, 24, 32, 48$ and 64 is depicted in Fig. 4. As indicated, q_{EA} increases as temperature decreases. However, the two systems exhibit different trends. For lattices with ANOB $= 3$, the finite value of q_{EA} at lower temperature systematically decreases as system sizes increases. This is the indication of the absence of SG phase in the thermodynamic limit. In contrast, for lattices with ANOB $= 4.0$, the finite value of the parameter at lower temperature remains, indicating the existence of SG phase in the thermodynamic limit. There is a significant increase of q_{EA} for system sizes $L = 16, 24$ and 32, which is a clear sign for the existence of SG phase in the rewired square lattices for ANOB=4.0. Although for larger sizes, i.e., $L = 48$ and 64, the value of the parameter at lower temperature decreases as system sizes increases, similar to the trend for the case of ANOB $= 3.0$, they are not systematic. The conflicting trends shown by the smaller and larger sizes for lattices with ANOB=4.0, signify the importance of dimensionality (coordination number) for the existence of ordered phase. The lattices with smaller sizes, i.e., $L < 48$, inherit the small-world behavior, which corresponds to properties of larger dimensional lattices. Therefore, although there is a sign for the existence of SG phase in the rewired square lattices for ANOB=4.0, we believe that a more convincing indication should be clearer for lattices with ANOB>4.0. Further study with larger ANOB, presently probed, is required.

4. Summary and Conclusion

We have studied antiferromagnetic Ising model on rewired square lattices with connectivity density (ANOB) 3.0 and 4.0; obtained by adding one and two extra bonds to each site of the original square lattice. The first added bond is constrained to connect to one of the next-nearest neighbors while the second one to one of the next-next-nearest neighbors. We use Replica Exchange Monte Carlo method which is considered very powerful in dealing with randomly frustrated systems such as SGs. Several physical quantities such as energy, specific heat and overlapping SG order parameter were calculated.

An indication SG phase transition is observed in the lattices with ANOB $= 4.0$ and not for ANOB $= 3.0$. This is indicated by a clear crossing point in the plot of overlapping parameter q_{EA}. There is no crossing point in the plot of the quantity for ANOB=3.0. A clear crossing point in the plot of q_{EA} for smaller sizes is indicative to the effect of small-world behavior of the lattice. In search for a clear indication of SG phase, we need to study lattices with connectivity density, ANOB> 4.0. A more comprehnsive study of system with larger fractional value of ANOB is in progress. The results will be reported elsewhere.

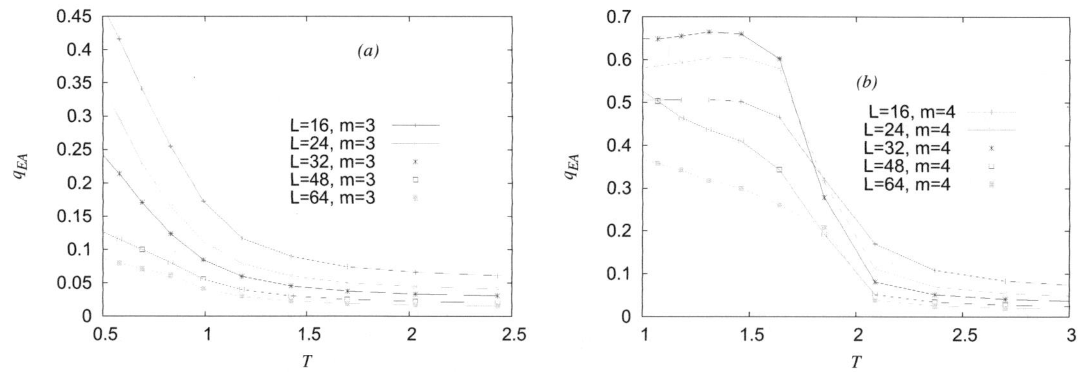

Figure 4. Temperature dependence of the overlapping order parameter for lattices with (a) ANOB $= 3.0$ and (b) 4.0 for various system sizes.

Acknowledgment
The author wishes to thank Bansawang BJ, D. Tahir and A. W. Sandvik for fruitful discussions. The computation of this work was performed using parallel computing facility in the Department of Physics Hasanuddin University and the HPC facility of Indonesian Institute of Science. The work is supported by PUPT Research Grant 2017 of the Indonesian Ministry of Research, Technology and Tertiary Education.

[1] B. Idzikowski *et al.,* Europhys. Lett.**45**, (1999) 714.
[2] F. Barahona, J. Phys. A15, (1982) 3241.
[3] N. Kawashima, J. Phys. Soc. Jpn., Vol. 75, No. 7 (2006) 073002.
[4] H. Nishimori, Statistical Physics of Spin Glasses and Information Processing, Oxford Univ. Press, (2001).
[5] V. Cannella and J. A. Mydosh, Phys. Rev. B6, (1972), 4220.
[6] V. Dupuis *et al.,* Phys. Rev. B64, (2001) 174204.
[7] S. V. Isakov, I. N. Zintchenko, T. F. Rønnow and M. Troyer, Comput. Phys. Commun., 192, (2015) 265.
[8] J. H. Holland, Adaptation in Natural and Artificial Systems, Univ. of Michigan Press, Michigan. (1975).
[9] M. Bartolozzi, T. Surungan, D.B. Leinweber and A.G. Williams, Phys. Rev. B73, (2006) 224419.
[10] C. P. Herrero, Phys. Rev. E **77**, (2008) 041102.
[11] T. Surungan, F. P. Zen, and A.G. Williams, JPCS 640, (2015) 012005.
[12] T. Surungan, JPCS 759, 012012, (2016).
[13] A.-L. Barabási and R. Albert, Science 286, (1999) 509.
[14] T. Surungan, Bansawang BJ, D. Tahir, AIP Conf. Proc. 1719, (2016) 030006.
[15] T. Surungan, Y. Okabe, and Y. Tomita, J. Phys. A37, (2004) 4219.
[16] T. Surungan, Y. Okabe and N. Kawashima, Phys. Rev B77, (2008) 214401.
[17] K. Hukushima and K. Nemoto, J. Phys. Soc. Japan 65, (1996) 1863.
[18] S. F. Edwards and P. W. Anderson, J. Phys. F: Metal Phys. **5**, (1975) 965.
[19] H. Yoshizawa *et. al.,* Physica B, **180**, (1992) 209.
[20] S. Wakimoto *et. al.,* Phys. Rev. B62, (2000) 3547.
[21] H. Martinho *et. al.,* Phys. Rev. B64, (2001) 024408.

IOP Conf. Series: Journal of Physics: Conf. Series **988** (2018) 012002 doi:10.1088/1742-6596/988/1/012002

Neutrino mass sum-rule

Asan Damanik

Department of Physics Education, Sanata Dharma University
Kampus III USD Paingan, Maguwoharjo, Sleman, Yogyakarta, Indonesia

E-mail: asandamanik11@gmail.com

Abstract. Neutrino mass sum-rele is a very important research subject from theoretical side because neutrino oscillation experiment only gave us two squared-mass differences and three mixing angles. We review neutrino mass sum-rule in literature that have been reported by many authors and discuss its phenomenological implications.

1. Introduction

As we have already knew from the Standard Model of Particle Physics especially electroweak interaction model based on $SU(2)_L \times U(1)_Y$ gauge group, it is not possibble to obtain a neutrino mass term in the Lagrangian of electroweak interaction when neurino to be put as a Dirac particle. But, if neutrino is a Majorana particle, then we can have a mass term in the Lagrangian which is given by

$$L = \frac{1}{2}\nu_L C^{-1} M \nu_L + h,c. \tag{1}$$

where ν_L contains a three left-handed neutrino fileds, C is the charge conjugation matrix, and M is the Majorana mass matrix.

It was a very long time, before the neutrino oscillation phenomena was reported by the Superkamiokande collaboration in 1998 [1], neutrino mass is assummed to be zero or approximately zero. Unfortunately, the neutrino oscillations experiments only gave us the squared-mass difference between two neutrino flavors that undergo oscillations during its propagation in vacuum (two sqared-mass differences), and three mixing angles that cannot be used to determined the absolute value of neutrino mass and its hierarchy. Another type of experiment that can be used to detect and determine the neutrino mass is neutrinoless double beta decay experiment. But, neutrinoless double beta decay experiment only give us upper bound of Majorana neutrino mass which is known as effective Majorana mass $\langle m_{ee} \rangle$. Thus, in order to determine neurino masses by using the experimental data as an input, we should seek another way or relation as an additional parameter that can be used to determine neutrino masses. One of the relation that can be used to help us in determining the absolute value of neutrino mass is the relation that link all three neutrino masses which is know as *neutrino mass sum-rule*.

In this paper, we review neutrino mass sum-rule that have already reported by many authors and discuss its phenomenological implications on neutrino masses, mass hierarchy, and effective Majorana mass. The paper is organized as follow: in section 2 we review neutrino mass sum-rule that have already reported by many authors; in section 3 we discuss the penomenological

IOP Conf. Series: Journal of Physics: Conf. Series **988** (2018) 012002 doi:10.1088/1742-6596/988/1/012002

implications of the neutrino mass sum rule on neutrino mass and effective Majorana mass especially for normal hierarchy. Finally, the section 4 is devoted to conclusions.

2. Brief review of neutrino mass sum-rule

Neutrino mass sum rule is a relation among the neutrino masses m_1, m_2, m_3 which are known to be very small and it very useful for determining i. e. the hierarchy of neutrino mass whether it normal or inverted hierarchy, the absolute values of neutrino masses, and the effective neutrino mass $|m_{ee}|$ as measured in neutrinoless double beta decay when we use the experimental data of neutrino oscillation as input. The importance of neutrino mass sum rule relation has already been stressed as well in e. g. Refs [2, 3, 4]. The neutrino mass sum rule, which can be obtained from several flavor models based on non-Abelian discrete symmetries, can be classified into four neutrino mass sum rules as one can reads in Ref. [6]

$$\chi m_2 + \xi m_3 = m_1, \tag{2}$$

$$\frac{\chi}{m_2} + \frac{\xi}{m_3} = \frac{1}{m_1}, \tag{3}$$

$$\chi\sqrt{m_2} + \xi\sqrt{m_3} = \sqrt{m_1}, \tag{4}$$

$$\frac{\chi}{\sqrt{m_2}} + \frac{\xi}{\sqrt{m_3}} = \frac{1}{\sqrt{m_1}}, \tag{5}$$

where χ and ξ are model dependent complex constants. A sample of various neutrino mass sum rule and the groups generating them are summarized in [7] and a summary table of the present neutrino mass sum rule in literatures can be found in [4, 5]. As pointed out in [8] that the fisrt three mass sum rule, a classifcation of all models predicting tribimaximal (TBM) mixing which generates mass relations similar to the first three sum rule, but the last case is a completely new case.

By referring to Ref. [4, 5], in literature, we have known that there are twelve neutrino mass sum-rule according to the general mass sum-rule that can be parameterized as follow

$$\begin{aligned} &s(m_1, m_2, m_3, c_1, c_2, \phi_1, \phi_2, d, \Delta_{\chi 13}, \Delta_{\chi 23}) \equiv \\ &\quad c_1\left(m_1 e^{-i\phi_1}\right)^d e^{i\Delta_{\chi 13}} + c_2\left(m_2 e^{-i\phi_2}\right)^d e^{i\Delta_{\chi 23}} + m_3^d = 0, \end{aligned} \tag{6}$$

where ϕ_1 and ϕ_2 are Majorana phases, and the quantities $c_1, c_2, d, \Delta_{\chi 13}$, and $\Delta_{\chi 23}$ are the parameters that characterize the sum-rule.

If we put $\phi_1 = \phi_2 = 0$ and $\Delta_{\chi 13} = \Delta_{\chi 23} = 0$, then Eq. (6) reads

$$c_1\left(m_1\right)^d + c_2\left(m_2\right)^d + m_3^d = 0. \tag{7}$$

It is apparent from Eq. (7) the neutrino mass sum-rule of Eq. (2) is easily obtained when we put $d = 1$, $c_1 = -\chi$, and $c_2 = -\xi$. To obtain Eq. (3) from Eq. (7) we should put $d = -1$, $c_1 = -\chi$, and $c_2 = -\xi$, and Eq. (4) is reproduced when we put $d = \frac{1}{2}$, $c_1 = -\chi$, and $c_2 = -\xi$. Finally, Eq. (5) will be obtained when we put $d = -\frac{1}{2}$, $c_1 = -\chi$, and $c_2 = -\xi$ into Eq. (7). It is an important task to explain why there are four possible values of parameter d and to decide what are the feasible value of d which is in agreement with the experimental data. Another question is why there are twelve values for χ and ξ that make possible twelve type of neutrino mass sum-rule as one can find in literature (see Table 1). In this paper, we do not explain or answer the above questions, but we only evaluate and discuss the phenomenological implications of four types of neutrino mass sum-rule as shown in Eqs. (2)-(5).

According to the experimental result of neutrino oscillation that the experimen only measure the squared mass difference, not absolute value of neutrino masses i. e. $\Delta m_{21}^2 > 0$ and $\Delta m_{31}^2 > 0$

IOP Conf. Series: Journal of Physics: Conf. Series **988** (2018) 012002 doi:10.1088/1742-6596/988/1/012002

or $\Delta m^2_{31} < 0$, then we can have two possible hierarchies of neutrino mass i. e. normal hierarchy (NH) when $\Delta m^2_{21} > 0$ and $\Delta m^2_{31} > 0$ and inverted hierarchy (IH) when $\Delta m^2_{21} > 0$ and $\Delta m^2_{31} < 0$.

Meanwhlie, neutrinoless double beta decay experiment only give us an upper bound of the effective Majorana mass $\langle m_{ee} \rangle$. The effective Majorana mass is given by

$$\langle m_{ee} \rangle = \left| \Sigma V^2_{ei} m_i \right|, \tag{8}$$

where V_{ei} is the i-th element of the first row of neutrino mixing matrix and m_i is the i-th of the neutrino mass.

Table 1. Possible neutrino mass sum-rule [7].

d	Type	Neurino mass sum-rule	Group
1	1	$m_1 + m_2 = m_3$	$A_4, A_5, S_4, \Delta(54)$
	2	$m_1 + m_3 = 2m_2$	S_4
	3	$2m_2 + m_3 = m_1$	A_4, S_4, T', T_7
	4	$m_1 + m_2 = 2m_3$	S_4
	5	$m_1 + \frac{\sqrt{3}+1}{2} m_3 = \frac{\sqrt{3}-1}{2} m_2$	A'_5
-1	6	$m_1^{-1} + m_2^{-1} = m_3^{-1}$	A_4, S_4, A_5
	7	$2m_2^{-1} + m_3^{-1} = m_1^{-1}$	A_4, T'
	8	$m_1^{-1} + m_3^{-1} = 2m_2^{-1}$	A_4, T'
	9	$m_3^{-1} \pm i m_2^{-1} = m_1^{-1}$	$\Delta(96)$
1/2	10	$\sqrt{m_1} - \sqrt{m_3} = 2\sqrt{m_2}$	$A_4 \times Z_2$
	11	$\sqrt{m_1} + \sqrt{m_3} = 2\sqrt{m_2}$	A_4
-1/2	12	$m_1^{-1/2} + m_2^{-1/2} = 2m_3^{-1/2}$	S_4

From Tabel 1, we can see that the A_4 symmetry is the most widely used to describe neutrino mass sum-rule. Since the A_4 symmetry is the most widwly used as the underlying symmetry of the neutrino mass sum-rule, therefore we only evaluate the neutrino mass sum-rules that can be described by A_4 symmetry and other symmetries that can proceed the same mass sum-rule. Thus, for the next section we only evaluate and discuss neutrino mass sum-rules of type 1 and 3 in Table 1 because those types of neutrino mass sum-rule are the most general mass sum-rule according to the number of underlying symmetries that can be used to describe it.

3. Phenomenological implications of neutrino mass sum-rule

As stated in the previously, we only evaluate the most general neutrino masses in Table 1 (type 1 and 3) prediction on neutrino masses when we use the data of neutrino oscillations as input.

3.1. Neutrino mass sum-rule of type 1

The neutrino mass sum-rule of type 1 reads

$$m_1 + m_2 = m_3. \tag{9}$$

From Eq. (9) we can have the following relation

$$m_2^2 + 2m_1 m_2 - \Delta m^2_{31} = 0, \tag{10}$$

where $\Delta m_{31}^2 = m_3^2 - m_1^2$. After doing a little algebra, the Eq. (10) proceed

$$m_2 = -2m_1 + \sqrt{4m_1^2 + \Delta m_{31}^2}. \tag{11}$$

We have also another squared-mass difference

$$\Delta m_{21}^2 = m_2^2 - m_1^2, \tag{12}$$

which can be measured in neutrino oscillation experiment. By inserting Eq. (11) into Eq. (12) and solving it to find m_1, then we have

$$m_1 = \frac{\sqrt{-105\Delta m_{21}^2 - 15\Delta m_{31}^2 + 60\sqrt{4\Delta m_{21}^4 - \Delta m_{21}^2\Delta m_{31}^2 + \Delta m_{31}^4}}}{15}, \tag{13}$$

or

$$m_1 = \frac{\sqrt{-105\Delta m_{21}^2 - 15\Delta m_{31}^2 - 60\sqrt{4\Delta m_{21}^4 - \Delta m_{21}^2\Delta m_{31}^2 + \Delta m_{31}^4}}}{15}. \tag{14}$$

By inserting the central values of squared-mass difference [9]

$$\Delta m_{21}^2 = 7.59 \times 10^{-5}\ \mathrm{eV}^2, \tag{15}$$
$$\Delta m_{31}^2 = 2.46 \times 10^{-3}\ \mathrm{eV}^2, \text{for NH} \tag{16}$$
$$\Delta m_{31}^2 = -2.36 \times 10^{-3}\ \mathrm{eV}^2, \text{for IH} \tag{17}$$

into Eq. (13), and Eqs. (11) and (9), then we have

$$m_1 = 0.021158\ \mathrm{eV},\ m_2 = 0.022881\ \mathrm{eV},\ m_3 = 0.04404\ \mathrm{eV}, \tag{18}$$

for normal hierarchy (NH): $|m_1| < |m_2| < |m_3|$, and

$$m_1 = 0.027615\ \mathrm{eV},\ m_2 = -0.028956\ \mathrm{eV},\ m_3 = -0.001342\ \mathrm{eV}, \tag{19}$$

for inverted hierarchy (IH): $|m_3| < |m_1| < |m_2|$.

If we use Eq. (14) to determine m_1, m_2 from Eq. (11), and m_3 from Eq. (9)then we have the hierarchy of neutrino masses as follow

$$|m_1| < |m_3| < |m_2|, \tag{20}$$
$$|m_2| < |m_2| < |m_1|, \tag{21}$$

when $\Delta m_{31}^2 > 0$, and

$$|m_2| < |m_1| < |m_3|, \tag{22}$$
$$|m_1| < |m_3| < |m_2|, \tag{23}$$

when $\Delta m_{31}^2 < 0$. Thus, only neutrino mass of Eq. (13) with neutrino mass sum-rule of type 1 can predict the hierarchy of neutrino mass in agreement with the experimental data of neutrino oscillations. Plot of effective Majorana mass as function of the lightest neutrino mass for neutrino mass sum rule of type 1 (red line for NH and green line for IH) is displayed in Figure 1.

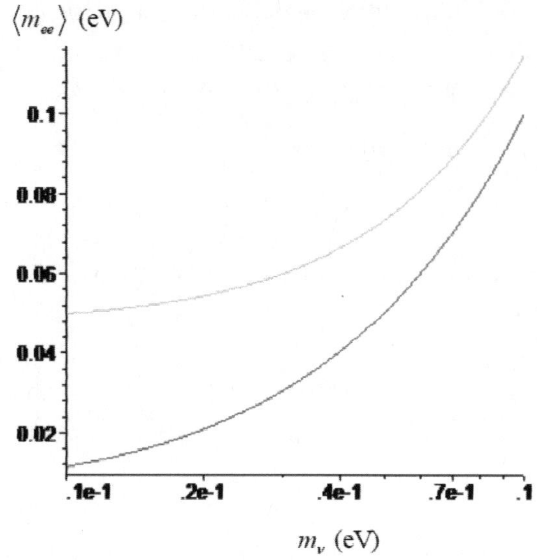

Figure 1. Plot of $\langle m_{ee} \rangle$ as function of lightest m_ν for sum-rule of type 1

3.2. Neutrino mass sum-rule oftType 3

As shown in Table 1, the neutrino mass sum-rule of type 3 reads

$$2m_2 + m_3 = m_1. \tag{24}$$

From neutrino mass sum-rule of Eq. (24) we can have

$$4m_2^2 + 4m_3m_2 + \Delta m_{31}^2 = 0, \tag{25}$$

which then proceed

$$m_2 = -\frac{\Delta m_{31}^2 - 4\Delta m_{32}^2}{2\sqrt{-\Delta m_{31}^2 + 4\Delta m_{32}^2}}, \tag{26}$$

or

$$m_2 = \frac{\Delta m_{31}^2 - 4\Delta m_{32}^2}{2\sqrt{-\Delta m_{31}^2 + 4\Delta m_{32}^2}}. \tag{27}$$

It is apparent from Eqs. (26) and (27) that neutrino mass m_2 is only as function of squared-mass differences Δm_{31}^2 and Δm_{32}^2. Since we need squared-mass difference m_{32}^2 to find out the value of m_2, then we can use the advantage of defenition squared-mass differences m_{21}^2 and m_{31}^2 which then proceeds

$$\Delta m_{32}^2 = \Delta m_{31}^2 - \Delta m_{21}^2. \tag{28}$$

By applying the same above procedure in determining the neutrino masses, from Eq. (26) we have neutrino masses for NH as follow

$$m_1 = 0.015869 \text{ eV}, \ m_2 = 0.018103 \text{ eV}, \ m_3 = 0.052075 \text{ eV}, \tag{29}$$

and when using Eq. (27) we have

$$m_1 = 0.015869 \text{ eV}, \; m_2 = 0.033972 \text{ eV}, \; m_3 = -0.052075 \text{ eV}, \tag{30}$$

which is consistent with normal hierarchy:$|m_1| < |m_2| < |m_3|$.

Meanwhile, for IH, by using Eq. (27) we have neutrino masses

$$m_1 = 0.018790i \text{ eV}, \; m_2 = -0.016648i \text{ eV}, \; m_3 = 0.052087i \text{ eV}, \tag{31}$$

and when using Eq. (27) we have

$$m_1 = 0.018790i \text{ eV}, \; m_2 = 0.035439i \text{ eV}, \; m_3 = -0.052087i \text{ eV}. \tag{32}$$

Both the obtained neutrino masses in Eqs. (31) and (32) are incosistent with the inverted hierarchy. Thus, we can olnly use the neutrino mass m_2 of Eq. (26) and neutrino mass sum-rule of type 3 to predict the correct hierarchy of neutrino mass. Plot of effective Majorana mass as function of the lightest neutrino mass for neutrino mass sum rule of type 3 (only allowed NH) is displayed in Figure 2.

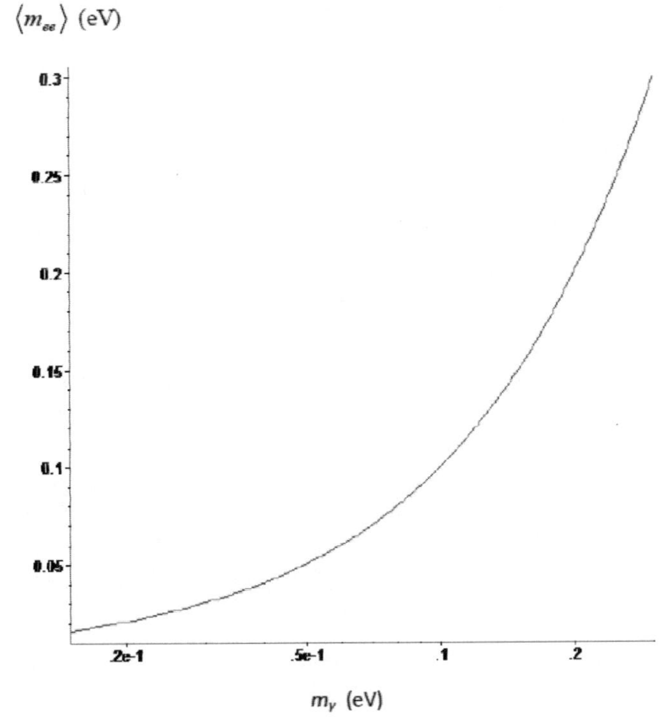

Figure 2. Plot of $\langle m_{ee} \rangle$ as function of m_ν for sum-rule of type 3

4. Conclusions

We have briefly review neutrino mass sum-rule that can be read in lietratures and we found there 12 type of neutrino mass sum-rule that can be derived from various symmetries. The most widely symmetry is A_4 symmetry. Based on the widely used symmetry that can be applied to derive the neutrino mass sum-rule, we choose two tyoes neutrino mass sum-rule that have

IOP Conf. Series: Journal of Physics: Conf. Series **988** (2018) 012002 doi:10.1088/1742-6596/988/1/012002

already been reported in literature i. e. type 1 and type 3. Whwn we evaluate the predictions of those both neutrino mass sum-rules on neutrino masses and its hierarchy by using the advantages of neutrino oiscillations data as input, we find that the neutrino mass sum-rule of type 1 can predict neutrino mass hierarchy both in normal hierarchy and inverted hierarcy. Meanwhile, the neutrino mass sum-rule of type 3 can only predict neutrino mass hierarchy in normal hierarchy.

Acknowledgement
Author would like to thank the Organizer for a nice hospitality during the CTPNP 2017 conference. This researh is partly supported by Hibah Penelitian Internal LPPM USD with contract No. 070/Penel./LPPM-USD/IV/2017.

References
[1] Kamiokande Collab.(Y. Fukuda *et al.*),*Phys. Rev. Lett.* **81**, 1562 (1998).
[2] M. -C. Chen and S. F. King, JHEP **0906**, 072 (2009).
[3] J. Barry and W. Rodejohann, Phys. Rev. **D81**, 093002 (2010).
[4] M. Spinrath, arXiv: 1609.07708 [hep-ph].
[5] J. Gehrlein and M. Spinrath, arXiv: 1704.02371 [hep-ph]
[6] L. Dorame, S. Morisi, E. Peinado, and J. W. F. Valle, Nucl. Phys. **B861**, 529 (2012).
[7] S. F. King, A. Merle, and A. J. Stuart, JHEP, **1312**, 005 (2013).
[8] A. D. Rojas, J. Phys.: Conf. Series **485**, 012045 (2014).
[9] M. C. Gonzales-Garcia, M. Maltoni and J. Salvado, JHEP **04**, 056 (2010).

A pedagogical derivation of the matrix element method in particle physics data analysis

Suharyo Sumowidagdo

Research Center for Physics, Indonesian Institute of Sciences Serpong, Banten, INDONESIA

E-mail: Suharyo.Sumowidagdo@lipi.go.id

Abstract. The matrix element method provides a direct connection between the underlying theory of particle physics processes and detector-level physical observables. I am presenting a pedagogically-oriented derivation of the matrix element method, drawing from elementary concepts in probability theory, statistics, and the process of experimental measurements. The level of treatment should be suitable for beginning research student in phenomenology and experimental high energy physics.

The matrix element method as known in high energy physics, was first proposed by Kondo [1, 2] and later by Dalitz and Goldstein [3, 4]. The method provides direct connection between the underlying theory of particle physics processes and physical observables measured by the particle detector. An early successful application of the method is the re-measurement of the top quark mass using D0 Run I data [5] which yields the most precise single-measurement from Tevatron Run I data.

Despite the method has been proposed almost 30 years ago, an elementary and pedagogical derivation of the method is not easy to find. Discussions and treatments of the method are scattered in original research papers, preprints, and dissertation. Furthermore, most of the existing treatments are written assuming the reader has a good knowledge of maximum likelihood estimation method. The treatment in this article is intended to be an entry-level treatment of the subject, after reading which beginning research students are ready to read more specialized treatments on the subject.

1. The concept of likelihood for single-variable probability distributions

To understand the matrix element method, one need to first understand the concept of likelihood as it is defined in statistics and probability theory. The treatment in this article is not meant to be comprehensive and rigorous, but rather a quick introduction into the key necessary concepts. A more elaborate discussions on probability theory, statistics, and likelihood can be found elsewhere, e.g. in the text by James [6].

Consider a set of measurements of a continuous variable x. Each measurement is performed under controlled, identical experimental situation. The outcome of each measurement is independent of the outcomes of other measurements. Then x is a random variable. The full set of measurements is defined as the *sample*. A single measurement in the sample is named an *event*.

If the continuous variable x is hypothesized to follow a probability distribution function $P(x)$, then the likelihood of an individual measurement which has value x_1 is defined as

$$\mathcal{L}(x_1) = P(x_1). \tag{1}$$

The likelihood that a sample which has n events to have measured a set of values $\{x_1, x_2, \ldots, x_n\}$ is the product of all likelihoods of the individual events

$$\mathcal{L}(\{x_1, x_2, \ldots, x_n\}) = P(x_1) \times P(x_2) \times \ldots \times P(x_n) = \prod_{i=1}^{n} P(x_i). \tag{2}$$

For computational purpose, the negative log-likelihood of a sample is preferred, which is defined as the negative of the sum of logarithms of all likelihoods

$$-\ln \mathcal{L}(\{x_1, x_2, \ldots, x_n\}) = -(\ln P(x_1) + \ln P(x_2) + \ldots + \ln P(x_n)) = -\sum_{i=1}^{n} \ln P(x_i). \tag{3}$$

Since the logarithm of a function is monotonically increasing function, the extremum points (maxima and minima) of the log-likelihood function will be located at the same extremum points of the likelihood function.

It is important to note that the probability distribution function $P(x)$ is *a prior*. In the context of experimental physics, $P(x)$ represent the theory to be tested; or used as theoretical input, in the experiment.

2. Multivariate probability distributions

The discussion in the previous section can be generalized for multivariate probability distributions. Consider a simultaneous measurement of k independent variable in each event. Thus x in the previous section becomes a vector $\mathbf{x} = (x_1, x_2, x_3, \ldots, x_k)$. If all the measured variables are independent of one another, the generalization of Equations (1), (2), and (3) are

$$\mathcal{L}(\mathbf{x}_1) = P(\mathbf{x}_1), \tag{4}$$

$$\mathcal{L}(\{\mathbf{x}_1, \mathbf{x}_2, \ldots, \mathbf{x}_n\}) = P(\mathbf{x}_1) \times P(\mathbf{x}_2) \times \ldots \times P(\mathbf{x}_n) = \prod_{i=1}^{n} P(\mathbf{x}_i), \tag{5}$$

$$-\ln \mathcal{L}(\{\mathbf{x}_1, \mathbf{x}_2, \ldots, \mathbf{x}_n\}) = -(\ln P(\mathbf{x}_1) + \ln P(\mathbf{x}_2) + \ldots + \ln P(\mathbf{x}_n)) = -\sum_{i=1}^{n} \ln P(\mathbf{x}_i). \tag{6}$$

In physics process, $\mathbf{x}$ is an independent set of variables which uniquelly characterizes each event. In the context of scattering process, $\mathbf{x}$ is the set of final-state particles' momenta, subject to the constraint of energy-momentum conservation.

3. Parameter estimation using likelihood

In using the likelihood for parameter estimation, the likelihood for an event is expressed as a function of a set of free parameters $\boldsymbol{\mu}$, $\mathcal{L} = \mathcal{L}(\mathbf{x}, \boldsymbol{\mu})$. The negative log-likelihood to observe a sample of n events is expressed as

$$-\ln \mathcal{L}(\{\mathbf{x}_1, \mathbf{x}_2, \ldots, \mathbf{x}_n\}, \boldsymbol{\mu}) = -\sum_{i=1}^{n} \ln \mathcal{L}(\mathbf{x}_i, \boldsymbol{\mu}). \tag{7}$$

Minimizing the negative log-likelihood over the parameter space of $\boldsymbol{\mu}$ will yield an estimator for $\boldsymbol{\mu}$, $\tilde{\boldsymbol{\mu}}$. Assuming the central limit theorem applies, the likelihood for a sample will have a Gaussian shape,

$$\mathcal{L} \propto \exp\left(-\frac{1}{2}(\boldsymbol{\mu} - \tilde{\boldsymbol{\mu}})^{\mathrm{T}} \boldsymbol{\Sigma}^{-1} (\boldsymbol{\mu} - \tilde{\boldsymbol{\mu}})\right) \tag{8}$$

IOP Conf. Series: Journal of Physics: Conf. Series **988** (2018) 012003 doi:10.1088/1742-6596/988/1/012003

in the neighbourhood of $\tilde{\boldsymbol{\mu}}$. The region bounded by the surface where the value of $-\ln\mathcal{L}$ is higher by 0.5 relative to value of $-\ln\mathcal{L}$ evaluated at $\tilde{\boldsymbol{\mu}}$ is considered to be the one-sigma confidence region of the parameters.

If there is only one parameter to be estimated, μ, the likelihood will have an univariate Gaussian shape,

$$\mathcal{L} \propto \exp\left(-\frac{(\mu-\tilde{\mu})^2}{2\sigma}\right). \tag{9}$$

The negative log-likelihood $-\ln\mathcal{L}$ curve has the shape of a concave upward parabola. The estimator of μ, $\tilde{\mu}$, is the value of μ correspond to the minimum of this curve. The one-sigma confidence interval of μ is take between the points where the value of the $-\ln\mathcal{L}$ is higher by 0.5 relative to the value of $-\ln\mathcal{L}$ evaluated at $\tilde{\mu}$

4. Hypothesis testing using likelihood

In using likelihood for hypothesis testing, two likelihoods are calculated each using two different hypotheses. The first likelihood is calculated using the *null hypothesis* H_0 as prior. The null hypothesis is a proposition which is expected to be true before the hypothesis testing is performed.

$$\mathcal{L}_0 = \mathcal{L}_0(\mathbf{x}|H_0) \tag{10}$$

The second likelihood is calculated using the *alternative hypothesis* H_c as prior. The alternative hypothesis is a proposition which is not expected to be true before the hypothesis testing. Depends on the results of hypothesis testing, the alternative hypothesis may or may not be accepted as alternative to the null hypothesis.

$$\mathcal{L}_c = \mathcal{L}_c(\mathbf{x}|H_c) \tag{11}$$

The Neyman-Pearson lemma [7] states that the ratio between the two likelihoods, Λ, is the optimal discriminating variable between the two hypothesis,

$$\Lambda = \frac{\mathcal{L}_0}{\mathcal{L}_c} \tag{12}$$

The hypothesis testing is done by comparing the likelihood ratio with a pre-determined value c, which corresponds to a certain significance level. If it is found that the likelihood ratio is less than c,

$$\Lambda < c \tag{13}$$

then the null hypothesis is rejected and the alternative hypothesis is accepted. If the likelihood ratio is equal or larger than c, the null hypothesis is not rejected.

5. Derivation of the matrix element method

In experimental particle physics, one observed the outcomes from collisions of pre-determined initial condition. Each individual collisions is the equivalent of *event* in the previous discussions. For each event, there are multiple quantities measured by the particle detector, in the forms of final particles' momenta. Thus the likelihood of an event is a function of more than one variable.

Consider a physics process $v_1 + v_2 \to p_1 + p_2 + \ldots + p_k$, where $v_1, v_2, p_1, p_2, \ldots, p_k$ are all four-momenta. We shall use the notation $\mathbf{v}_i$ to indicate the initial set of four-momenta $\{v_1, v_2\}$, and the notation $\mathbf{p}_f$ to indicate the final set of four-momenta $\{p_1, p_2, \ldots, p_k\}$. The *parton-level* differential cross-section for this process can be expressed as

$$d\sigma\left(\mathbf{p}_f|\mathbf{v}_i\right) = \frac{1}{F(\mathbf{v}_i)}(2\pi)^4|\mathcal{M}(\mathbf{v}_i \to \mathbf{p}_f)|^2\delta^4\left(v_1 + v_2 - \sum_{f=1}^{f=k} p_f\right)\prod_{f=1}^{f=k}\frac{d^3p_f}{(2\pi)^3(2E_f)} \tag{14}$$

IOP Conf. Series: Journal of Physics: Conf. Series **988** (2018) 012003 doi:10.1088/1742-6596/988/1/012003

as stated in standard field theory textbooks (e.g. [8]). Here F is the flux factor which may includes convolution over the parton distribution functions, and $\mathcal{M}$ is the *matrix element* or *Lorentz-invariant scattering amplitude* of the process. In practice, the detector will smear a parton-level four-momentum p_f into detector-level four-momentum q_f. To express the differential cross-section as a functions of $\mathbf{q}_f$ instead of $\mathbf{p}_f$, we must unsmear each individual detector-level four-momenta q_f, i.e., express q_f as a convolution of p_f.

$$q_f = \int W\left(q_f|p_f\right) d^3p_f \tag{15}$$

An important assumption about the transfer function warrants a discussion. The transfer function for a parton-level four-momenta p_i into detector-level four-momenta q_i is assumpted to be independent of the other parton-level and detector-level four-momenta. In other words, the complete transfer function W can be factorized into individual transfer function W_i

$$W(\mathbf{q}_f|\mathbf{p}_f) = W_1(q_1|p_1) \times W_2(q_2|p_2) \times W_3(q_3|p_3) \times \ldots \times W_k(q_k|p_k) \tag{16}$$

Thus the detector-level differential cross-section, $d\sigma(\mathbf{q}_f|\mathbf{v}_i)$ is an integral of the all possible parton-level differential cross-section $d\sigma(\mathbf{p}_f|\mathbf{v}_i)$ weighed by the transfer function $W(\mathbf{q}_f|\mathbf{p}_f)$.

$$\begin{aligned} d\sigma\left(\mathbf{q}_f|\mathbf{v}_i\right) &= \int W\left(\mathbf{q}_f|\mathbf{p}_f\right) d\sigma(\mathbf{p}_f|\mathbf{v}_i) \\ &= \int W\left(\mathbf{q}_f|\mathbf{p}_f\right) \frac{1}{F(v_1,v_2)}(2\pi)^4|\mathcal{M}(\mathbf{v}_i \to \mathbf{p}_f)|^2\delta^4\left(v_1+v_2-\sum_{f=1}^{f=k} p_f\right)\prod_{f=1}^{f=k}\frac{d^3p_f}{(2\pi)^3(2E_f)} \end{aligned} \tag{17}$$

We can normalized the expression for detector-level differential cross-section with the detector-level total cross-section, σ_{t} to obtain a normalized probability density (thus likelihood).

$$\mathcal{L}\left(\mathbf{q}_f|\mathbf{v}_i\right) = \frac{1}{\sigma_{\mathrm{t}}}\int W\left(\mathbf{q}_f|\mathbf{p}_f\right) \frac{1}{F(\mathbf{v}_i)}(2\pi)^4|\mathcal{M}(\mathbf{v}_i \to \mathbf{p}_f)|^2\delta^4\left(v_1+v_2-\sum_{f=1}^{f=k} p_f\right)\prod_{f=1}^{f=k}\frac{d^3p_f}{(2\pi)^3(2E_f)} \tag{18}$$

Equation 18 is the matrix element method. It express the likelihood to observe an event with detector-level momenta $\mathbf{q}_f$, given a prior of the underlying physics process in the matrix element $\mathcal{M}$ and detector effect W. The matrix element $\mathcal{M}$ contains the input theoretical model (*a priori*) and encapsulates the physics of the process under study. In practice sometimes only the dominant processes are calculated, at the lowest-order Feynman diagram. The four-dimensional δ-function represents momentum conservation and will reduce the number of independent variables to $3k-4$, where k is the number of parton-level particles in the final state. The product term in the Equation 18 represents the final state phase space of parton-level four-momenta $\mathbf{p}_f$ after unsmearing of detector effects. To compute the likelihood, the equation must be integrated (numerically) over this phase space. Often, the integration of the phase is the most difficult task in using the matrix element method.

In many applications, there are more than one physics process that can happen and compatible with a given sets of initial and final state particles. One of the process usually is identified as *signal process*, while the others are identified as *background processes*. For example, in the study of top quark-antitop quark pair production in the $e^+e^-\nu_e\bar{\nu}_e b\bar{b}q\bar{q}$ channel at the LHC, the signal process is $pp \to t\bar{t} \to W^+bW^-\bar{b} \to e^+e^-\nu_e\bar{\nu}_e b\bar{b}q\bar{q}$ while the dominant background process are $pp \to Zb\bar{b} \to e^+e^-b\bar{b}$ and $pp \to Zjj \to e^+e^-jj$, where jj are either gluon or light-quark jets. If this case, then the likelihood for each event will be a sum of the likelihood for

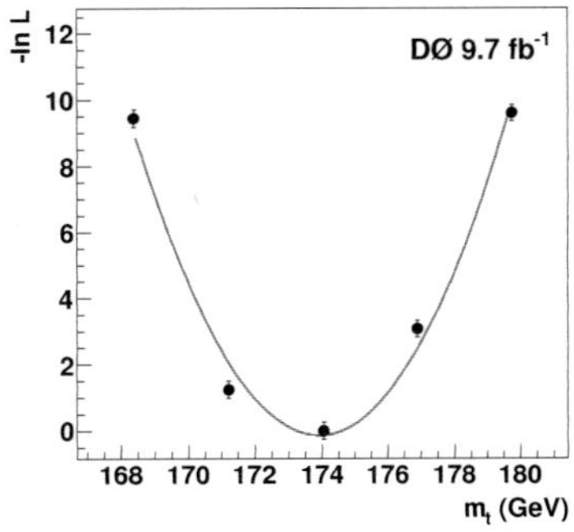

Figure 1. The likelihood evaluated as a function of five top quark mass values, and the fit of parabolic curve to the negative log-likelihood curve.

each processes, weighted by the fraction of each process in the data sample.

$$\mathcal{L} = \sum_c f_c \mathcal{L}_c \tag{19}$$

Here f_c is the fraction of events coming from process c in the data sample. The determination of f_c can be done either independently from, or simultaneously with, the application of matrix element method. In measurements, the fractions should be determined independently. In searches, the fraction can be determined simultaneously using the likelihood output from the matrix element method.

6. Example: Top quark mass measurement

A recent results is the measurement of top quark mass in the dilepton channel by D0 Collaboration [9]. The likelihood used in the measurement is

$$P(\mathbf{q}_f|f_{t\bar{t}}, m_t) = f_{t\bar{t}} P_{t\bar{t}}(\mathbf{q}_f|m_t) + (1 - f_{t\bar{t}}) P_{bkg}(\mathbf{q}_f) \tag{20}$$

A comprehensive treatment of the application of the matrix element method in measurements of top quark mass can be found in [10].

The fraction of $t\bar{t}$ events in the data sample is $f_{t\bar{t}}$, the fraction of background events is $f_{bkg} = 1 - f_{t\bar{t}}$. The probability for signal events, $P_{t\bar{t}}$, is a function of the top quark mass, while the probability for background events, P_{bkg}, is independent on the value of top quark mass. The signal probability is taken to be the quark fusion process $q\bar{q} \to t\bar{t} \to W^+bW^-\bar{b} \to \ell^+\ell^-\nu_\ell\bar{\nu}_\ell b\bar{b}q\bar{q}$, while the background matrix element is taken to be the $Z\gamma^* \to \ell^+\ell^- + 2$ jets process. The gluon fusion $t\bar{t}$ production process and the $Z\gamma^* \to \ell^+\ell^- b\bar{b}$ are both neglected due to their cross sections are relative smaller to the included processes. Both the signal and background matrix elements are calculated at the leading order.

The sample likelihood is calculated as a function of several top quark mass values. To correct for bias, the method is calibrated using $t\bar{t}$ Monte Carlo samples generated at the same set of top quark mass values used in the calculation of the sample likelihood. A parabolic curve (corresponding to Gaussian shape of the likelihood) is then fitted to negative log-likelihood curve, as shown in Figure 1.

7. Example: Spin correlation in top quark-antiquark production

The top quark decays through weak interactions before it hadronizes, thus its spin state information would be transferred cleanly to its decay products. Furthermore, due to its very

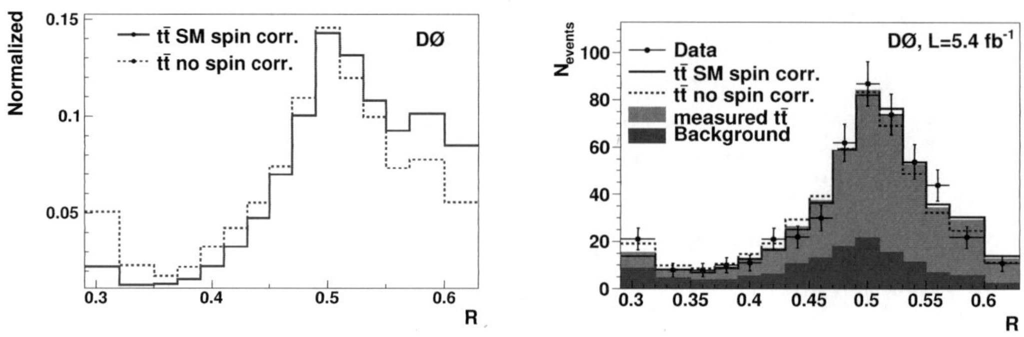

Figure 2. Left figure is the fit templates for $t\bar{t}$ with spin correlation (solid line) and without spin correlation (dotted line). Right figure is the results of the fit to data.

short-lifetime, top quark doesn't experience spin flips after it is produced. In top quark-antiquark production thus there is correlation information between the spin of the top quark and the spin of the top antiquark [11].

The D0 Collaboration has measured the fraction of $t\bar{t}$ with spin correlation using the matrix element method [12]. The details of the method was elaborated in this reference [13]. In the analysis, a variable $\mathcal{R}$, which is related to the likelihood ratio, is introduced with the definition

$$\mathcal{R} = \frac{P_{\text{signal}}(\text{correlated})}{P_{\text{signal}}(\text{correlated}) + P_{\text{signal}}(\text{uncorrelated})} \tag{21}$$

A binned fit of the $\mathcal{R}$ variable in data is performed, using templates from Monte Carlo (simulated) sample of $t\bar{t}$ events with and without spin correlation, to determine the fraction of $t\bar{t}$ events in data with spin correlation. Figure 2 shows the templates used, and the results of fitting the templates to data.

8. Acknowledgments

The author would like to thank D0 Collaboration for granting permission to use figures from their publications [9, 12] in this article.

References

[1] Kondo K 1988 *J. Phys. Soc. Jap.* **57** 4126–4140
[2] Kondo K 1991 *J. Phys. Soc. Jap.* **60** 836–844
[3] Dalitz R H and Goldstein G R 1992 *Phys. Rev.* **D45** 1531–1543
[4] Goldstein G R, Sliwa K and Dalitz R H 1993 *Phys. Rev.* **D47** 967–972 (*Preprint* hep-ph/9205246)
[5] Abazov V M *et al.* (D0 Collaboration) 2004 *Nature* **429** 638–642 (*Preprint* hep-ex/0406031)
[6] James F 2006 *Statistical methods in experimental physics* (World Scientific) ISBN 9789812567956
[7] Neyman, J and Pearson, Egon S 1933 *Phil. Trans. Roy. Soc. Lond.* **A231** 289–337
[8] Peskin M E and Schroeder D V 1995 *An Introduction to Quantum Field Theory* (Perseus Books Publishing)
[9] Abazov V M *et al.* (D0) 2016 *Phys. Rev.* **D94** 032004 (*Preprint* 1606.02814)
[10] Fiedler F, Grohsjean A, Haefner P and Schieferdecker P 2010 *Nucl. Instrum. Meth.* **A624** 203–218 (*Preprint* 1003.1316)
[11] Barger V D, Ohnemus J and Phillips R J N 1989 *Int. J. Mod. Phys.* **A4** 617
[12] Abazov V M *et al.* (D0) 2011 *Phys. Rev. Lett.* **107** 032001 (*Preprint* 1104.5194)
[13] Melnikov K and Schulze M 2011 *Phys. Lett.* **B700** 17–20 (*Preprint* 1103.2122)

Simulating The Residual Three-body Force in a Small System

Tony Sumaryada[1,]
[1]Theoretical Physics DivisionDepartment of Physics, Bogor Agricultural University, Jalan Meranti Kampus IPB Dramaga Bogor 16680, Indonesia

E-mail: tsumaryada@ipb.ac.id

Abstract. The effect of residual three-body force on the small system was investigated in this paper. A one-level and two-level system were chosen as the model to study the feature of three-body force due to their simplicity and lower symmetry. In both systems, we found a consistent result, that the inclusion of three-body force will enhance and accelerate the pairing condensation, which is shown by a higher pairing gap and lower energy as compared to the system with only two-body pairing force. Some interesting results on the behavior of the residual three-body force were also presented here.

Keywords: three-body force, pairing gap, small system, BCS theory

.

1. Introduction

Pairing correlations are one of the most common phenomena in the universe. It occurs (or predicted to occur naturally) everywhere, ranging from tiny quarks, nucleus, atom, material, or even in the astronomical body like neutron stars. It is just natural that interest to study the pairing correlation and all aspect related to that phenomena is always high. Since its first publication in 1957, BCS theory of superconductivity [1] continuously become the main reference for the scientist working in superconductivity, superfluidity, and pairing phenomena. Similar theory but using other approaches like Bogouliubov transformation and Landau theory of liquid also enrich our understanding in the pairing related research. The BCS theory initially was designed to explain the pairing phenomena namely electron's pairing in condensed matter/solid state, but soon was realized to be applicable to another system like atomic nuclei [2]. As long as we have a high-density state and lower temperature, the pairing condensation phenomena must naturally (spontaneously) be occurred. The idea of using BCS type equation into a three-body force in the nuclear and mesoscopic system, for example can be found in reference [3]. Some applications in the realistic nuclear problem can be seen in reference [4]. Some mathematical/spin algebra reviews and critics have also been done by other [5].

The mesoscopic or small system is a system that comprised of few particles up to thousands of particles. This small system is a perfect playground for the scientist to test and play around with the theory since it is small enough for quantum phenomena to occur yet adequate enough to be treated statistically.

In this paper, the effect of residual three-body force (R3BF) in the small system was investigated by using a BCS framework. The one-level and two-level systems were chosen as a model to study the general features and properties of the R3BF due to their simplicity and a straightforward demonstration of the effect.

2. Methods

The pairing interaction between two bodies resulted in the formation of pairing state which has lower energy compared to the normal (unpaired) state. This difference of energy expressed in a famous BCS pairing gap equation [1] as shown in Equation 1.

$$\Delta_k = \frac{1}{2}\sum_{k'>0} \frac{\Delta_k}{\sqrt{\varepsilon_k^2 + \Delta_k^2}} V_{kk'} \tag{1}$$

where Δ_k denotes the pairing gap at state k, and ε_k denotes the single particle energy at state k and $V_{kk'}$ denotes the interaction matrix between state k and k'. This equation can be solved numerically using a self-consistency method. In the BCS framework, the R3BF is formulated as :

$$\Delta_1' = \Omega_2 \frac{\Delta_2'}{4E_2'} \left(G_{12} - g_{12;3}\langle n_3\rangle\right) \tag{2}$$

With Δ_1 denotes the pairing gap of three-body force, while Ω_2 denotes the occupation number of the state, G_{12} denotes the two-body pairing constant, while $g_{12;3}$ denotes the three-body pairing constant. Detail of the above formula derivation can be found in [6]. The main features of the R3BF are (1) it is a sequential pairing in which the two-body interaction occurred first and create a single condensate before making another interaction with the incoming third particle, (2) the R3BF is a density-dependent interaction and becomes more significant at very high-density state.

The most important parameter in this study is the pairing gap enhancement which is expressed by :

$$\Delta_{enhc}(g_{12;3}) = \left|\Delta_{3B}(g_{12;3}) - \Delta_{2B}\right| \tag{3}$$

while the pairing energy enhancement is calculated by :

$$E_{enhc}(g_{12;3}) = \left|E_{3B}(g_{12;3}) - E_{2B}\right| \tag{4}$$

where Δ_{2B} and E_{2B} is the two-body pairing gap and the total energy of two-body pairing system (at $g_{12;3}=0$) respectively. The parameters of $\Delta_{3B}(g_{12;3})$ and $E_{3B}(g_{12;3})$ denote the three-body pairing gap and the total energy of the three-body force system at a particular three-body constant $g_{12;3}$. The effect of R3BF in this paper will be shown in terms of pairing enhancement parameter, such as a pairing gap enhancement and a pairing energy enhancement. All calculations were performed using the FORTRAN77 program under a Fedora 14 Linux platform.

3. Result and Discussions

In the one-level model, we assume that all particles have the same energy and lying on a single degenerate system. The half occupied system was chosen in this paper since it has the maximum pairing gap compared to other configurations. In Fig.1a, the pairing gap as a function of a number of particles (N) for even N is shown. The calculation was started from N = 4 since there is no three-body force working on N = 2. It was found that when $g_{12;3}$ is increased (black line is indicating the absence of three-body force, while the red line indicating three-body constant $g_{12;3}$ = 0.0050) the pairing gap is increased as well. A bigger pairing gap means a more stable pairing phase is formed separated from the normal phase. In the absence of three-body force, the maximum pairing gap occurs when the system is half occupied ($\omega = \Omega/2$), where Ω is the total occupation number and ω is the half occupation number of the energy level. The introduction of R3BF will shift the maximum pairing gap to the right side ($N > \omega$), depending on the value of three-body constant $g_{12;3}$.

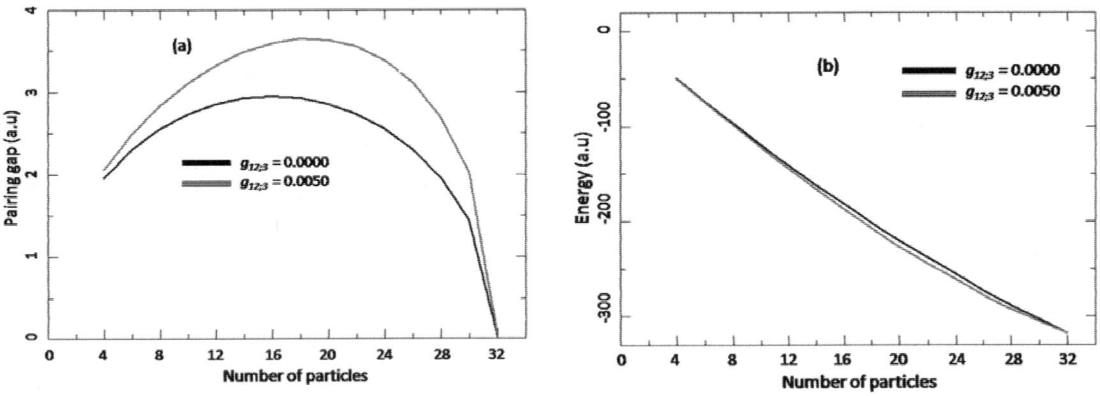

Figure 1. (a) Pairing gap vs Number of Particle (N) of one level system for $g_{12;3}$=0.000 (without three-body force) and $g_{12;3}$=0.0050. (b) The total energy of the system for $g_{12;3}$=0.000 and $g_{12;3}$=0.0050.

For the energy of the system, the presence of three-body force is again enhancing the pairing interaction and condensation as indicated by a lower energy of the system. As an illustration, let's take a look at N = 20 particles; when there is no three-body force present, the total energy of the system is -220.216 a.u (in arbitrary unit), but when $g_{12;3}$ is 0.0050 the energy becomes -226.513 a.u. The behavior of the total energy of the system as a function of a number of particles is shown in Fig 1b. Similar to pairing gap, the effect of the three-body force also vanishes when the system is fully occupied. The addition of R3BF to a one-level system will increase the energy gap and enhance the stability of pairing state by lowering its energy.

It is also interesting to study the effect of a number of particles to the pairing (gap) enhancement due to a three-body constant. In Fig. 2, the pairing enhancement as a function of three-body constant were shown for two different half occupied systems, N = 16 and N = 50. The pairing (gap) enhancement of R3BF as compared to a regular two-body pairing shows a linear dependent to the number of particles N and this signifies the importance of R3BF at a higher density as seen in Fig.2.

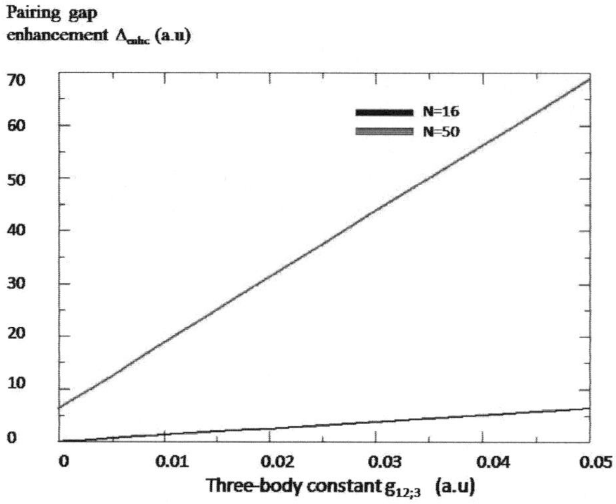

Figure 2. The pairing gap enhancement of R3BF as compared to regular two-body pairing gap for half occupied one level systems with N=16 and N=50 particles.

In the case of the two-level system, there is a critical two-body pairing constant (G_c) in which the pairing effect start to occur. The introduction of the three-body constant will lower the critical pairing strength and accelerates the occurrence of pairing condensate as can be seen in Fig.3a, where the critical two-body pairing constant G_c which initially 0.47 (no three-body constant presence) now shifted to 0.08 when $g_{12;3}$=0.002. At $g_{12;3}$=0.003, G_c is even vanishes which means that without any two-body pairing strength, the pairing condensation will still occur at particular three-body constant. The same effect can also be seen in term of the total energy of the system as shown in Fig 3b. In general, the presence of R3BF accelerates the pairing condensation by shifting the critical two-body pairing strength into lower values or even annihilate the two-body pairing constant.

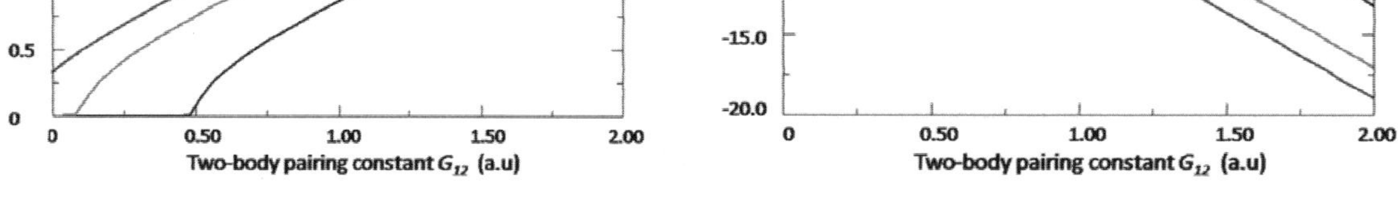

Figure 3. The pairing acceleration process due to the introduction of R3BF which lowering the critical two-body pairing constant as seen from the (a) pairing gap (b) total energy of N=20 particles of a half occupied two-level system.

The effect of R3BF on the two-level system is also shown in Fig. 4. The pairing gap enhancement is behaving non-linearly when the applied two-body pairing constant below its critical point G_c ($G< G_c$), and tend to be constant at the region above its critical point ($G> G_c$). The application of small three-body constant at the $G> G_c$ region does not introduce any new behavior as shown by its monotonic behavior of pairing gap and total energy enhancement versus G_{12} in Fig 4a and 4b.

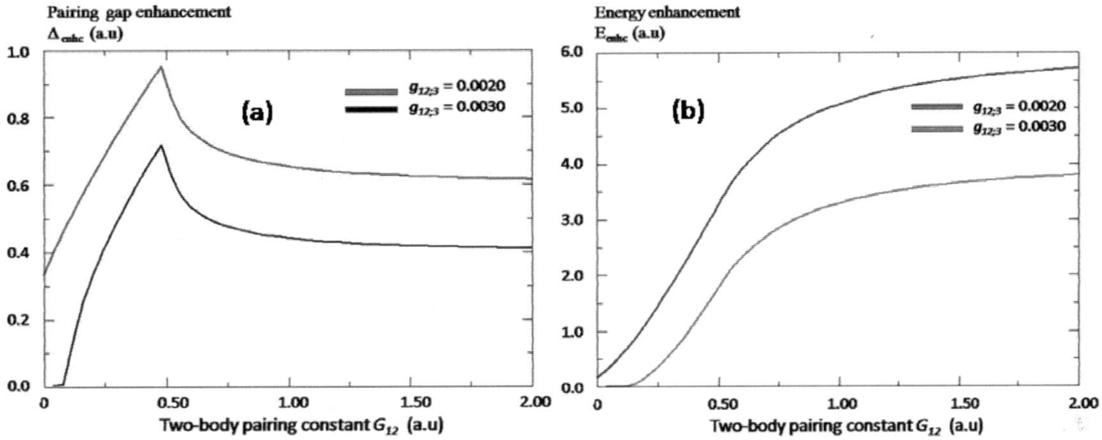

Figure 4. (a) The non-linear behavior of pairing gap enhancement, and (b) the total energy of the system at various three-body constant in the two-level system.

4. Conclusion

The effect of the residual three-body force in the small system has been investigated using a BCS framework by introducing a three-body constant into a pairing gap equation. The results have shown that the R3BF, in general, enhances the pairing condensation by widening the gap and lowering the energy of the system. The pairing enhancement is density dependent and increases as the number of particles in the half occupied degenerate system increased. In a two-level system, the effect of R3BF is non-linear for $G< G_c$ and tend to be constant at $G>G_c$. The monotonic behavior (constant pairing enhancement) of R3BF at $G>G_c$ shows that the three-body constant seems to act as an "amplifier" for the two-body pairing interaction as shown in the pairing gap and energy enhancement and does not introduce any new-type of interaction.

Acknowledgements

The author expresess his gratitude to A. Volya and V. Zelevinsky for introducing the three-body problems in nuclear structure. The author also expresses his gratitude to Bogor Agricultural University for the computational support, and Frostburg State University, Committee of Faculty Development and Sabbatical Leave for partial funding in this project (the year 2010).

References

[1] Bardeen J, Cooper L N and Schrieffer J R 1957 *Phys. Rev.* **108** 1175
[2] S. T. Belyaev, Mat. Fys. Medd. Dan. Vid. Selsk. 31, No. 11 (1959)
[3] Zelevisnky V., Yad. Fiz. (Phys. At. Nucl.), 72 (2009), 1107
[4] Volya A., EPL, 86 (2009) 52001
[5] Isacker P. Van, Talmi. I , EPL, 90 (2010) 32001
[6] Sumaryada T I 2007 Pairing Correlations and Phase Transitions in Mesoscopic Systems Dissertation FSU.

An outline of cellular automaton universe via cosmological KdV equation

V Christianto[1], F Smarandache[2] and Y Umniyati[3]

[1] Malang Institute of Agriculture, Malang, Indonesia.
[2] Department of Mathematics and Sciences, University of New Mexico, Gallup, USA
[3] Physics Laboratory, Department of Mechatronics, Swiss German University, Jalur Sutera barat No. 15, Alam Sutera, Tangerang, Indonesia

E-mail: victorchristianto@gmail.com; Founder of http://www.ketindo.com

Abstract. It has been known for long time that the cosmic sound wave was there since the early epoch of the Universe. Signatures of its existence are abound. However, such a sound wave model of cosmology is rarely developed fully into a complete framework. This paper can be considered as our *second* attempt towards such a complete description of the Universe based on soliton wave solution of cosmological KdV equation. Then we advance further this KdV equation by virtue of Cellular Automaton method to solve the PDEs. We submit wholeheartedly Robert Kuruczs hypothesis that *Big Bang should be replaced with a finite cellular automaton universe with no expansion* [4][5]. Nonetheless, we are fully aware that our model is far from being complete, but it appears the proposed cellular automaton model of the Universe is very close in spirit to what Konrad Zuse envisaged long time ago. It is our hope that the new proposed method can be verified with observation data. But we admit that our model is still in its infancy, more researches are needed to fill all the missing details.

1. Introduction

Konrad Zuse is probably the first scholar who imagine a *Computing Universe.* In recent years, there are a few researchers who suggest similar vision in terms of cellular automata, for example Stephen Wolfram, Gerardus t Hooft, and Robert Kurucz from Harvard Smithsonian of Astrophysics. Nonetheless, it seems that there is no existing model which can be connected with a nonlinear PDE of the Universe. In this paper, we try to offer some working CA models based on the KdV equation, which can be modelled and solved using computer algebra packages such as Mathematica.

Meanwhile, Korteweg-de Vries (KdV) equation is a non-linear wave equation plays a fundamental role in diverse branches of mathematical and theoretical physics. Its significance to cosmology has been discussed by a number of authors, such as Rosu and recently Lidsey [3][7]. It is suggested that the KdV equation arises in a number of important scenarios, including inflationary cosmology etc. Analogies can be drawn between cosmic dynamics and the propagation of the solitonic wave solution to the equation, whereby quantities such as the speed and amplitude profile of the wave can be identified with cosmological parameters such as the spectral index of the density perturbation spectrum and the energy density of the universe.

2. Cosmological KdV equation

The Korteweg-de Vries (KdV) equation is the completely integrable, third-order, non-linear partial differential equation (PDE) [3]:

$$\partial_t u + \partial_x^3 u + \frac{3}{u_0} u \partial_x u = 0, \tag{1}$$

where $u = u(x,t), \partial_t = \frac{\partial}{\partial t}, \partial_x^3 = \frac{\partial^3}{\partial x^3}$, etc., u_0 is a constant and (x,t) represent space and time coordinates, respectively. This equation was originally derived within the context of small-amplitude, non-linear water wave theory and it is well known that it admits a solitonic wave solution of the form

$$u = u_0 \lambda^2 \sec h^2 \left[\frac{\lambda(x - \lambda^2 t)}{2} \right], \tag{2}$$

where the constant $\frac{\lambda}{2}$ represents the wavenumber of the soliton. The KdV soliton is characterized by the property that its speed and amplitude are proportional to the square of the wavenumber.

Rosu [7] and also Lidsey [3] both have considered some cosmological applications of KdV equation. We will consider here one application in inflationary universe model. It can be shown that Friedmann equation after some steps which have been discussed in [3], yields to an equation which takes the form of (2), as follows:

$$H^2(\phi) = H_0^2 \lambda^2 \sec h^2 \left[\frac{\lambda A}{2} \right], \tag{3}$$

where

$$A = \frac{\sqrt{8\pi}}{m_p} \phi. \tag{4}$$

Therefore, it appears quite reasonable to consider this equation as originated from certain cosmological KdV physics.

3. Cellular automata model of KdV equation: Towards cellular automaton universe

There are several methods to consider discretization of KdV equation into cellular automata models. Here we briefly discuss only few methods:

(i) Based on paper by [11], KdV equation can be written as a conservation law:

$$\frac{\partial u}{\partial t} + \frac{\partial}{\partial x} \left(-\frac{u^2}{2} - \frac{\partial^2 u}{\partial x^2} \right) = 0. \tag{5}$$

It follows that, after the simplest discretization, we obtain the cellular automata:

$$u_j(t+1) = u_j(t)[u_{j+1}(t) - u_j(t)] + u_{j+2}(t) - u_{j+1}(t) - u_{j-1}(t). \tag{6}$$

Thus $\Sigma_{j=0}^{N-1} u_j(t)$ is not an invariant.

(ii) The discrete analogue of the KdV equation is known thanks to the pioneering work of Hirota. It has the form [15]:

$$\frac{1}{u_{l+1}^{t+1}} - \frac{1}{u_l^t} = \delta(u_{l+1}^t - u_l^{t+1}) \tag{7}$$

IOP Conf. Series: Journal of Physics: Conf. Series **988** (2018) 012005 doi:10.1088/1742-6596/988/1/012005

(iii) Another model was proposed by Tokihiro et al around twenty years ago. They suggested that an integrable discretization (differential-difference equation) of the KdV equation is the Lotka-Volterra equation [14]:

$$\frac{d}{dt}b_jt = b_j(t)[b_{j+t}(t) - b_{j-1}(t)] \tag{8}$$

In other words, it appears possible at least in theory to consider a Cellular Automaton-KdV Universe, based on discretization of the original KdV equation. Nonetheless, further analysis is required to study its potential applications.

4. Discussion

We advance further KdV equation by virtue to Cellular Automaton method to solve the PDEs. Here, we consider some mathematical methods to discretize the original KdV equation in order to be transformed into cellular automata cosmology models. As far as our knowledge, this approach is different from existing cellular cosmology, such as by Conrad Ranzan (www.cellularuniverse.com).

References

[1] Pain H J 2005 *The Physics of Vibrations and Waves* (New York: John Wiley and Sons)
[2] Hu W and White M 2004 *The Cosmic Symphony*, Scientific American 2004. URL: https://www.scientificamerican.com/author/wayne-hu-and-martin-white/
[3] Lidsey J E Cosmology and the Korteweg-de Vries equation, Phys. Rev. D. vol. 86, URL: https://journals.aps.org/prd/pdf/10.1103/PhysRevD.86.123523.
[4] Kurucz R L Radiatively-driven cosmology in the cellular automaton universe. Harvard-Smithsonian Center for Astrophysics, 60 Garden St, Cambridge, MA 02138. URL: http://kurucz.harvard.edu/papers/radcellular/radiativelycell.ps
[5] Kurucz R L Elementary physics in the cellular automaton universe. Harvard-Smithsonian Center for Astrophysics, 60 Garden St, Cambridge. URL: http://kurucz.harvard.edu/papers/elementary/elementary.ps
[6] Adiwidjaja H 1995 On the Korteweg-de Vries equation. CDS 141 Project, Caltech Univ. URL: www.cds.caltech.edu/ marsden/wiki/uploads/projects/geomech/Adiwidjaja1995.pdf
[7] Rosu H C 2002 *Mod. Phys. Lett.* A **17** 667
[8] Scott A C 2007 *The Nonlinear Universe* (Berlin: Springer-Verlag)
[9] Wolfram S 1994 *Cellular Automata and Complexity* (Addison-Wesley)
[10] Wolfram S. Cellular Automaton fluids I: basic theory. URL: www.stephenwolfram.com/publications/academic/cellular-automaton-fluids-theory.pdf
[11] Steeb W H and Hardy Y 2007 *Naturforsch* **62a** 567
[12] Yang X S and Young Y 1983 Cellular automata, PDEs, and pattern formation. Chapter in Handbook of Bioinspired Algorithms and Applications, edited by Stephan Olariu, Albert Y. Zomaya. (New York: Chapman and Hall/CRC Computer and Information Science Series, 2010)
[13] de Oliveira P P B and Verardo M 2014 *Math. J.* **16** url: http://www.mathematica-journal.com/2014/08/representing-families-of-cellular-automata-rules/
[14] Tokihiro T 1996 *Phys. Rev. Lett.* **76** 18
[15] Willox R 2010 *J. Phys. A: Math. Theor.* **43** 482003
[16] Kurucz R L 2000 The formation of life. Harvard-Smithsonian Center for Astrophysics, 60 Garden St, Cambridge, MA 02138
[17] Walker S I 2017 Origins of life: A problem for physics. *Reports on Progress in Physics, IOP* url: http://iopscience.iop.org/article/10.1088/1361-6633/aa7804/pdf, https://doi.org/10.1088/1361-6633/aa7804

IOP Conf. Series: Journal of Physics: Conf. Series **988** (2018) 012006 doi:10.1088/1742-6596/988/1/012006

Testing the new BPS method in some models of nonabelian magnetic monopole

I Prasetyo[1,2], **A N Atmaja**[2] **and H S Ramadhan**[1]

[1] Departemen Fisika, FMIPA, Universitas Indonesia, Depok 16424, Indonesia
[2] Research Center for Physics, Indonesian Institute of Sciences (LIPI), Kompleks PUSPIPTEK Serpong, Tangerang 15310, Indonesia

E-mail: ilham.prasetyo@sci.ui.ac.id

Abstract. The proposed method in [*Phys. Lett.* **B768** 351-358 (2017)], which can obtain BPS equations of some models of vortices, is used here to see whether it is still usable for some models of magnetic monopole. Other than the standard Yang-Mills-Higgs, here we report that the method is able to give us the BPS equations from two different magnetic monopole models.

1. Introduction

Monopole is a non-contractible defect whose second homotopy group over 2-sphere is nontrivial [1–4]. Magnetic monopole has received much interest since Dirac [5] proposed a certain vector potential in an abelian gauge theory. Generalized to a nonabelian one, 't Hooft [6] and Polyakov [7, 8] had shown that magnetic monopole arises from such non-abelian model. In their seminal paper, Prasad and Sommerfield [9] showed that the corresponding second-order field equations can have analytic solutions in certain limit. Later Bogomolny [10], based on classical energy stability, showed that the field equations can be reduced to the first-order ones. Further, he proved that such solutions are stable since their energy is globally minimum. The corresponding first-order equations and exact solutions are called the BPS (Bogomolny-Prasad-Sommerfield) equations and solutions, respectively.

For certain models, some techniques to obtain BPS equations had been presented, such as [11, 12] for recent proposals. Though the method proposed in [12] uses one of the results (i.e. BPS energy functional Q) from the On-Shell method proposed in [11] as its basic motivation, the former is noticeably easier to use. Here we apply such method to magnetic monopoles. Although both vortices and monopoles depend on scalar and gauge fields configuration, the former is Abelian while the latter in not. We show that the method can be applied to finding the BPS equations for certain noncanonical models of magnetic monopoles.

The discussion in this proceeding paper are as follows. We review the magnetic monopole from a nonabelian gauge theory in section 2, then we describe the method from [12] in section 3. Then we show the models whose BPS equations are able to be found using the method in the next three sections, i.e., standard Yang-Mills-Higgs, generalized Yang-Mills-Higgs, and Nakamula-Shiraishi DBI model [13] in section 4, 5, and 6, respectively. Lastly since it is a work in progress, some remarks and future projections are discussed in section 7.

2. Magnetic monopole from nonabelian gauge field

We begin with action in a flat $(1+3)$-dimensional spacetime whose metric is $\eta_{\mu\nu}$ with signature $(+,-,-,-)$ with the Lagrangian density having a standard form

$$\mathcal{L} = \frac{1}{2}\mathcal{D}_\mu\phi^a\mathcal{D}^\mu\phi^a - V(|\phi|) - \frac{1}{4}F^{a\mu\nu}F^a_{\mu\nu}, \tag{1}$$

with the components of the adjoint covariant derivative and the field strength tensor are defined respectively as $(a = 1,2,3)$

$$\mathcal{D}_\mu\phi^a = \partial_\mu\phi^a + e\epsilon^{abc}A^b_\mu\phi^c, \tag{2}$$

$$F^a_{\mu\nu} = \partial_\mu A^a_\nu - \partial_\nu A^a_\mu + e\epsilon^{abc}A^b_\mu A^c_\nu. \tag{3}$$

The scalar and Yang-Mills gauge field is written as $\phi = \phi^a T^a$ and $\mathcal{A} = \mathcal{A}_\mu dx^\mu = A^a_\mu T^a dx^\mu$, respectively, with $T^a \equiv \sigma^a/2$ denoting anti-Hermitian generators of $\mathfrak{su}(2)$ algebra, whose commutation relation is $[T^a, T^b] = i\epsilon^{abc}T^c$ and its trace is $\mathrm{tr}(T^aT^b) = \frac{1}{2}\delta^{ab}$. Here we employ the 't Hooft-Polyakov hedgehog ansatz

$$\phi^a = \eta f(r)\frac{x^a}{r}, \; A^a_0 = 0, \tag{4}$$

$$A^a_m = \frac{1-a(r)}{e}\epsilon^{amn}\frac{x^n}{r^2}, \tag{5}$$

where $x^a = (x,y,z)$ denotes Cartesian coordinate. In spherical coordinates (r,θ,φ), their components are

$$\phi = \eta f(\sin\theta(\cos\varphi T^1 + \sin\varphi T^2) + \cos\theta T^3), \tag{6}$$

$$\mathcal{A}_r = 0, \; \mathcal{A}_\theta = \frac{1-a}{e}(\sin\varphi T^1 - \cos\varphi T^2), \tag{7}$$

$$\mathcal{A}_\varphi = \frac{1-a}{e}\sin\theta(\cos\theta(\cos\varphi T^1 + \sin\varphi T^2) - \sin\theta T^3). \tag{8}$$

Writing them this way make it easier to obtain the components of the adjoint covariant derivative of the scalar field $\mathcal{D}_\mu\phi = \partial_\mu\phi - ie[\mathcal{A}_\mu, \phi]$ and the field strength tensor $\mathcal{F}_{\mu\nu} = \partial_\mu\mathcal{A}_\nu - \partial_\nu\mathcal{A}_\mu - ie[\mathcal{A}_\mu, \mathcal{A}_\nu]$, which are

$$\mathcal{D}_r\phi = \eta f'(r)(\sin\theta(\cos\varphi T^1 + \sin\varphi T^2) + \cos\theta T^3), \tag{9}$$

$$\mathcal{D}_\theta\phi = \eta a f(\cos\theta(\cos\varphi T^1 + \sin\varphi T^2) - \sin\theta T^3), \tag{10}$$

$$\mathcal{D}_\varphi\phi = \eta a f\sin\theta(-\sin\varphi T^1 + \cos\varphi T^2), \tag{11}$$

$$\mathcal{F}_{r\theta} = \frac{a'(r)}{e}(-\sin\varphi T^1 + \cos\varphi T^2), \tag{12}$$

$$\mathcal{F}_{r\varphi} = \frac{a'(r)}{e}\sin\theta(-\cos\theta(\cos\varphi T^1 + \sin\varphi T^2) + \sin\theta T^3), \tag{13}$$

$$\mathcal{F}_{\theta\varphi} = \frac{a^2-1}{e}\sin\theta(\sin\theta(\cos\varphi T^1 + \sin\varphi T^2) + \cos\theta T^3), \tag{14}$$

with prime denoting derivative of its variable, e.g., $f'(r) \equiv df/dr$, so we obtain

$$\begin{aligned}
\frac{1}{2}\mathcal{D}_\mu\phi^a\mathcal{D}^\mu\phi^a &= \text{tr}\ \mathcal{D}_\mu\phi\mathcal{D}^\mu\phi \\
&= \frac{\eta^{rr}(\eta f'(r))^2 + (\eta^{\theta\theta} + \eta^{\varphi\varphi}\sin^2\theta)(\eta a f)^2}{4}\ \text{tr}\begin{pmatrix}1 & 0\\ 0 & 1\end{pmatrix} \\
&= -\frac{(\eta f'(r))^2}{2} - \left(\frac{\eta a f}{r}\right)^2, \qquad (15)\\
\frac{1}{4}F^{a\mu\nu}F^a{}_{\mu\nu} &= \text{tr}\ \frac{1}{2}\mathcal{F}^{\mu\nu}\mathcal{F}{}_{\mu\nu} \\
&= \frac{\eta^{rr}(\eta^{\theta\theta} + \eta^{\varphi\varphi}\sin^2\theta)\left(\frac{a'(r)}{e}\right)^2 + \eta^{\theta\theta}\eta^{\varphi\varphi}\sin^2\theta\left(\frac{1-a(r)^2}{e}\right)^2}{4}\ \text{tr}\begin{pmatrix}1 & 0\\ 0 & 1\end{pmatrix} \\
&= \left(\frac{a'(r)}{er}\right)^2 + \frac{1}{2}\left(\frac{1-a^2}{er^2}\right)^2. \qquad (16)
\end{aligned}$$

Hence the Lagrangian density becomes

$$\begin{aligned}
\mathcal{L} &= \text{tr}\left(\mathcal{D}_\mu\phi\mathcal{D}^\mu\phi - \frac{1}{2}\mathcal{F}^{\mu\nu}\mathcal{F}{}_{\mu\nu}\right) - V(|\phi|) \\
&= -\frac{\eta^2 f'(r)^2}{2} - \left(\frac{\eta a f}{r}\right)^2 - V(f) - \left(\frac{a'(r)}{er}\right)^2 - \frac{1}{2}\left(\frac{a^2-1}{er^2}\right)^2. \qquad (17)
\end{aligned}$$

Using Bogomolny trick, i.e., when $V = 0$ we can arrange its static energy

$$\begin{aligned}
E_{\text{static}} &= -4\pi\int r^2 dr\left[-\frac{\eta^2 f'(r)^2}{2} - \left(\frac{\eta a f}{r}\right)^2 - \left(\frac{a'(r)}{er}\right)^2 - \frac{1}{2}\left(\frac{a^2-1}{er^2}\right)^2\right] \\
&= 4\pi\int r^2 dr\left[\frac{1}{2}\left\{\eta f'(r) \mp \frac{a^2-1}{er^2}\right\}^2 + \frac{\{a'(r) \mp e\eta a f\}^2}{e^2 r^2} \pm \frac{2\eta a'(r) a f}{er^2} \pm \frac{\eta f'(r)(a^2-1)}{er^2}\right] \\
&\geq \pm\frac{4\pi\eta}{e}\int r^2 dr\frac{\left[(a^2-1)f\right]'}{r^2} = \pm\frac{4\pi\eta}{e}(a^2-1)f\Big|_{r\to 0}^{r\to\infty} = \mp\frac{4\pi\eta}{e}. \qquad (18)
\end{aligned}$$

This lowest static energy $E_{\text{static}} = 4\pi\eta/e$ thus comes from the solution of the BPS equations

$$f'(r) = -\frac{a^2-1}{\eta e r^2}, \qquad (19)$$

$$a'(r) = -e\eta a f. \qquad (20)$$

In the following discussion, we simplify our notation by setting $e = 1 = \eta$.

3. The method

Here we follow the prescription on obtaining BPS equations in [12]. Any effective Lagrangian density, whose fields[1] are $\phi_i = (\phi_1, ..., \phi_N)$, can be added with some terms that do not change its Euler-Lagrange equations. These terms are called the boundary term(s) $\mathcal{L}_{\text{bd}}$ since their value is apparent only at the boundary of the space, while the Euler-Lagrange equations describe the system at all positions. Any static configuration has its Hamiltonian density equal to the

[1] They need not be scalars.

negative of its Lagrangian density, so its total energy is $E_{\text{static}} = -\int d^d x \mathcal{L}$, with d its spatial dimension. Then one can write the static energy as

$$E_{\text{static}} = \left(\int d^d x \sum_i \Phi_i(\phi_1, ..., \phi_N, \partial\phi_i; r) \right) + E_{\text{bd}}, \tag{21}$$

with Φ_i a set of positive-semidefinite functions that contains only one of its respective first derivative field and E_{BPS} is the lowest static energy available defined as $E_{\text{bd}} = -\int d^d x \mathcal{L}_{\text{bd}}$. The BPS limit is when the static energy is equal to its lowest bound $E_{\text{static}} = E_{\text{bd}}$, i.e., when $\Phi_i(\phi_1, ..., \phi_N, \partial\phi_i; r) = 0$. Then $\Phi_i(\phi_1, ..., \phi_N, \partial\phi_i; r) = 0$ are considered as the BPS equations and the boundary term in the Lagrangian density is then equal to BPS Lagrangian $\mathcal{L}_{\text{BPS}}$. Solutions from the BPS equations are solutions of the Euler-Lagrange equations at the BPS limit, thus we can write the Lagrangian density as

$$\mathcal{L} = -\left(\sum_i \Phi_i(\phi_1, ..., \phi_N, \partial\phi_i; r) \right) + \mathcal{L}_{\text{bd}}, \tag{22}$$

where at BPS limit, $\Phi_i(\phi_1, ..., \phi_N, \partial\phi_i; r) = 0$ and $\mathcal{L}_{\text{bd}} = \mathcal{L}_{\text{BPS}}$. Hence at the BPS limit, we can obtain the BPS equations by employing $\mathcal{L} - \mathcal{L}_{\text{BPS}} = 0$.

A result from the On-Shell method [11] is that, at least for vortices [12], the lowest possible static energy at the BPS limit can be defined as

$$E_{\text{bd}} = Q(r \to \infty) - Q(r \to 0) = \int_{r \to 0}^{r \to \infty} dQ, \tag{23}$$

where Q is called as BPS energy functional. This, as suspected in [12], depends only on the fields since the ansatz employed in the effective Lagrangian density has no explicit coordinate expression, such as the radial coordinate r. In this paper, the employed ansatz give us (9)-(14) that depend on r implicitly. Hence for general fields, Q is dependent only on the fields

$$Q \equiv \prod_{i=1}^{N} Q_i(\phi_i). \tag{24}$$

which makes

$$E_{\text{bd}} = \int \sum_i \frac{\partial Q}{\partial \phi_i} \frac{d\phi_i}{dr} dr, \tag{25}$$

and we can obtain $\mathcal{L}_{\text{BPS}}$.

Now after using a suitable Q to obtain $\mathcal{L}_{\text{BPS}}$ and then employing $\mathcal{L} - \mathcal{L}_{\text{BPS}} = 0$, we proceed to find all Φ_i. Notice that each Φ_i must be positive-semidefinite, hence each must has $\partial_r \phi_i$ as its roots, which if written as a set, the expression must be

$$\partial_r \phi_i = \left\{ F_i^{(1)}, F_i^{(2)}, ..., F_i^{(m)} \right\}, \tag{26}$$

with $F_i^{(k)} = F_i^{(k)}(\phi_1, ..., \phi_N, r)$ $(k = 1, ..., m)$ and $m =$ even number. If $\Phi_i = 0$ is a quadratic equation, hence $m = 2$ and each BPS solution ϕ_i will have a constraint $F_i^{(1)} = F_i^{(2)}$.

Generally, one can also find the BPS equations as follows. By seeing $\mathcal{L} - \mathcal{L}_{\text{BPS}} = 0$ as a polynomial equation of one of the fields, say $\partial_r \phi_1$, one can find its roots to be

$$\partial_r \phi_1 = \left\{ G_1^{(1)}, G_1^{(2)}, ..., G_1^{(m)} \right\}, \tag{27}$$

with $G_1^{(k)} = G_1^{(k)}(\phi_1, ..., \phi_N, \underline{\partial_r\phi_1}, \partial_r\phi_2, ..., \partial_r\phi_N, r)$ ($k = 1, ..., m$ and m = even number) with the underlined one is removed. If $m = 2$, then $G_1^{(1)} - G_1^{(2)} = 0$ is a polynomial function of another field $\partial_r\phi_i$ with $i > 1$. If the total of all $\partial_r\phi_i$ is $m = 2$, then by looping this steps until $i = N$ we arrive at the last equation $G_N^{(1)} - G_N^{(2)} = 0$. Since the model is valid for all r, we write $G_N^{(1)} - G_N^{(2)} = 0$ as terms with powers of r and we treat each term to be equal to zero. From each term, we can find each $Q_i(\phi_i)$ and the BPS equations $\partial_r\phi_i$ can then be found.

This method is shown to be adequate to obtain BPS equations for various models of vortices using BPS energy function $Q = 2\pi F(f)A(a)$ as shown in [12] and here we investigate whether the method is also adequate for some models of 't Hooft-Polyakov magnetic monopoles or not.

4. Standard Yang-Mills-Higgs model

Looking at the standard Lagrangian density (17) and its ansatz, we guess that (24) is true so we define the BPS energy function for magnetic monopole to be

$$Q = 4\pi F(f)A(a). \tag{28}$$

Since $E_{\text{BPS}} = -\int d^3x\mathcal{L}_{\text{BPS}} = -\int r^2 \sin\theta \; dr \; d\theta \; d\varphi \; \mathcal{L}_{\text{BPS}} = -4\pi \int r^2 dr \mathcal{L}_{\text{BPS}}$ we have

$$\mathcal{L}_{\text{BPS}} = -\frac{F'(f)A}{r^2} f' - \frac{FA'(a)}{r^2} a'. \tag{29}$$

From this point on, we will write $A'(a)$ as A' and $F'(f)$ as F' by keeping in mind that $A' = dA/da$ and $F' = dF/df$. We input this into $\mathcal{L} - \mathcal{L}_{\text{BPS}} = 0$ with $\mathcal{L}$ from (17) so we have

$$\frac{1}{2}f'^2 + \frac{a^2f^2}{r^2} + V + \frac{1}{r^2}a'^2 + \frac{(a^2-1)^2}{2r^4} - \frac{F'A}{r^2}f' - \frac{FA'}{r^2}a' = 0. \tag{30}$$

Now we choose explicitly the expression of both Φ_a and Φ_f. This can be done by choosing for Φ_a and Φ_f, respectively,

$$\frac{1}{r^2}a'^2 - \frac{FA'}{r^2}a' + \frac{a^2f^2}{r^2} = 0, \tag{31}$$

$$\frac{1}{2}f'^2 - \frac{F'A}{r^2}f' + V + \frac{(a^2-1)^2}{2r^4} = 0. \tag{32}$$

Each leads, respectively, to

$$a'_\pm = \frac{FA'}{2} \pm \sqrt{\left(\frac{FA'}{2}\right)^2 - a^2f^2}, \tag{33}$$

$$f'_\pm = \frac{F'A}{r^2} \pm \sqrt{\left(\frac{F'A}{r^2}\right)^2 - 2\left(V + \frac{(a^2-1)^2}{2r^4}\right)}. \tag{34}$$

Since we need Φ_a to be positive-semidefinite then $a'_+ = a'_-$ so

$$FA' = \pm 2af \tag{35}$$

which if we choose $F = f$ then $A = \pm(a^2 + k_A)$. We also need $f'_+ = f'_-$ so

$$2V = \left(\frac{(a^2+k_A)^2}{r^2}\right)^2 - \frac{(a^2-1)^2}{r^4} = 0. \tag{36}$$

IOP Conf. Series: Journal of Physics: Conf. Series **988** (2018) 012006 doi:10.1088/1742-6596/988/1/012006

The last equality is the condition for BPS equations to be found and it is true if $k_A = -1$. (If $k_A \neq -1$, then $V = V(a)$, which is not true.) Hence the BPS equations are

$$f' = \pm\frac{(a^2-1)}{r^2}, \tag{37}$$
$$a' = \pm af. \tag{38}$$

We can also find with another approach which in [12] is said to be more general than what we have described in section 3. We can treat $\mathcal{L} - \mathcal{L}_{\text{BPS}} = 0$ (30) as a quadratic equation of first derivative of one of the fields, where here we choose a'

$$\frac{1}{r^2}a'^2 - \frac{FA'}{r^2}a' + \left[\frac{1}{2}f'^2 + \frac{a^2f^2}{r^2} + V + \frac{(a^2-1)^2}{2r^4} - \frac{F'A}{r^2}f'\right] = 0, \tag{39}$$

hence the roots of a' are

$$a'_\pm = \frac{FA'}{2} \pm \sqrt{\frac{(FA')^2}{4} - r^2\left[\frac{1}{2}f'^2 + \frac{a^2f^2}{r^2} + V + \frac{(a^2-1)^2}{2r^4} - \frac{F'A}{r^2}f'\right]}. \tag{40}$$

Since we want $a'_+ = a'_-$, the square root must be zero, or after rearranging

$$\frac{r^2}{2}f'^2 - F'Af' - \frac{(FA')^2}{4} + a^2f^2 + V + \frac{(a^2-1)^2}{2r^2} = 0. \tag{41}$$

The roots are

$$f'_\pm = \frac{F'A}{r^2} \pm \sqrt{\frac{(F'A)^2}{r^4} - \left(-\frac{(FA')^2}{2r^2} + \frac{2a^2f^2}{r^2} + \frac{2V}{r^2} + \frac{(a^2-1)^2}{r^4}\right)}, \tag{42}$$

which again from $f'_+ = f'_-$ we need

$$\frac{(F'A)^2}{r^4} - \left(-\frac{(FA')^2}{2r^2} + \frac{2a^2f^2}{r^2} + \frac{2V}{r^2} + \frac{(a^2-1)^2}{r^4}\right) = 0. \tag{43}$$

This constraint can be solved by finding F and A through choosing each term with powers of r to be equal to zero, i.e.,

$$\frac{(F'A)^2}{r^4} - \frac{(a^2-1)^2}{r^4} = 0, \tag{44}$$
$$-\frac{(FA')^2}{2r^2} + \frac{2a^2f^2}{r^2} + \frac{2V}{r^2} = 0. \tag{45}$$

If we choose $F = f$ then from the former equation we have $A = \pm(a^2 + k_A)$ with k_A a constant, and the latter equation give us

$$2V = \frac{(\pm 2af)^2}{2} - 2a^2f^2 = 0, \tag{46}$$

which is the condition for BPS equations to be found. This leads to the same BPS equations as (37) and (38).

5. Generalized Yang-Mills-Higgs model

Now we consider the following Lagrangian density

$$\mathcal{L} = \text{tr}\left(w(|\phi|)\mathcal{D}_\mu\phi\mathcal{D}^\mu\phi - \frac{1}{2}G(|\phi|)\mathcal{F}^{\mu\nu}\mathcal{F}_{\mu\nu}\right) - V(|\phi|)$$
$$= -w(f)\left(\frac{f'(r)^2}{2} + \left(\frac{af}{r}\right)^2\right) - V(f) - G(f)\left(\left(\frac{a'(r)}{r}\right)^2 + \frac{1}{2}\left(\frac{a^2-1}{r^2}\right)^2\right), \quad (47)$$

that resembles the generalized Maxwell-Higgs model for vortices. Again by using the same $\mathcal{L}_{\text{BPS}}$ and choosing Φ_a and Φ_f, respectively, as

$$-w\left(\frac{af}{r}\right)^2 + \frac{FA'}{r^2}a' - G\left(\frac{a'(r)}{r}\right)^2 = 0, \quad (48)$$

$$-w\frac{f'^2}{2} + \frac{F'A}{r^2}f' - V(f) - \frac{G}{2}\left(\frac{a^2-1}{r^2}\right)^2 = 0. \quad (49)$$

The former equation leads to

$$a'_\pm = \frac{FA'}{2G} \pm \sqrt{\left(\frac{FA'}{2G}\right)^2 - \frac{w}{G}(af)^2}, \quad (50)$$

hence

$$FA' = \pm 2af\sqrt{wG}, \quad (51)$$

which gives us $F = f\sqrt{wG}$ and $A = \pm(a^2 + k_A)$. The latter equation leads to

$$f'_\pm = \frac{F'A}{wr^2} \pm \sqrt{\left(\frac{F'A}{wr^2}\right)^2 - \frac{2V}{w} - \frac{G}{w}\left(\frac{a^2-1}{r^2}\right)^2}, \quad (52)$$

which leads us to a constraint

$$\left(\frac{\partial(\sqrt{wG}f)}{\partial f}\right)^2 = \frac{wr^4}{(a^2+k_A)^2}\left[2V + G\left(\frac{a^2-1}{r^2}\right)^2\right]. \quad (53)$$

For the case of $V = 0$ and $k_A = -1$, we have $\left(\frac{\partial(\sqrt{wG}f)}{\partial f}\right)^2 = wG = const$. Hence the BPS equations are

$$a' = \pm\sqrt{\frac{w}{G}}af, \quad (54)$$

$$f' = \pm\sqrt{\frac{G}{w}}\frac{(a^2-1)}{r^2}. \quad (55)$$

Now we try using the second method as before. Arranging the equation to be

$$\frac{G}{r^2}a'^2 - \frac{FA'(a)}{r^2}a' + \left[\frac{w}{2}f'^2 - \frac{F'(f)A}{r^2}f' + \frac{wa^2f^2}{r^2} + V + \frac{G}{2}\frac{(a^2-1)^2}{r^4}\right] = 0, \quad (56)$$

we obtain

$$a'_\pm = \frac{FA'(a)}{2G} \pm \sqrt{\frac{F^2A'(a)^2}{4G^2} - \frac{r^2}{G}\left[\frac{w}{2}f'^2 - \frac{F'(f)A}{r^2}f' + \frac{wa^2f^2}{r^2} + V + \frac{G}{2}\frac{(a^2-1)^2}{r^4}\right]}. \quad (57)$$

Now to have $a'_+ = a'_-$, it must satisfy

$$\frac{w}{2}f'^2 - \frac{F'(f)A}{r^2}f' + \left[-\frac{F^2 A'(a)^2}{4Gr^2} + \frac{wa^2 f^2}{r^2} + V + \frac{G}{2}\frac{(a^2-1)^2}{r^4}\right] = 0, \tag{58}$$

whose roots are

$$f'_\pm = \frac{F'(f)A}{wr^2} \pm \sqrt{\frac{F'(f)^2 A^2}{w^2 r^4} + \frac{2}{w}\left[\frac{F^2 A'(a)^2}{4Gr^2} - \frac{wa^2 f^2}{r^2} - V - \frac{G}{2}\frac{(a^2-1)^2}{r^4}\right]}, \tag{59}$$

hence we have

$$\frac{F'(f)^2 A^2}{w^2 r^4} + \frac{F^2 A'(a)^2}{2wGr^2} - \frac{2a^2 f^2}{r^2} - \frac{2V}{w} - \frac{G}{w}\frac{(a^2-1)^2}{r^4} = 0 \tag{60}$$

to solve. Now we choose

$$\frac{F^2 A'(a)^2}{2wGr^2} - \frac{2a^2 f^2}{r^2} = 0 \tag{61}$$

so that $FA'(a) = \pm 2a\sqrt{wG}f$ and we choose $F = \sqrt{wG}f$ so $A = \pm(a^2 + k_A)$. The rest, i.e.,

$$\frac{F'(f)^2 A^2}{w^2 r^4} - \frac{2V}{w} - \frac{G}{w}\frac{(a^2-1)^2}{r^4} = 0 \tag{62}$$

leads to

$$V = \left[\frac{\partial(\sqrt{wG}f)}{\partial f}\right]^2 \frac{(a^2+k_A)^2}{2wr^4} - \frac{G(a^2-1)^2}{2r^4}. \tag{63}$$

Since V is a function of f, then $V = 0$, hence we have a constraint

$$\frac{\partial(\sqrt{wG}f)}{\partial f} = \pm\sqrt{\frac{wG(a^2-1)^2}{(a^2+k_A)^2}}. \tag{64}$$

Since the left hand side is a function only depending on f thus it require $k_A = -1$ which then makes $wG = const.$, which, without loss of generality, can be normalized to unity. Thus,

$$w = G^{-1}. \tag{65}$$

The resulting BPS equations are the same as (54) and (55).

6. Nakamula-Shiraishi DBI model

We begin with a Lagrangian density suggested in [13] that is guaranteed to have the standard BPS equations. Here we show that our method is still usable for this form of Lagrangian density. The Lagrangian density has the following form

$$\begin{aligned}\mathcal{L} =& -\beta^2 \mathrm{tr}\left(\sqrt{1+\frac{2}{\beta^2}\mathcal{D}_\mu\phi\mathcal{D}^\mu\phi}\sqrt{1+\frac{1}{\beta^2}\mathcal{F}_{\mu\nu}\mathcal{F}^{\mu\nu}} - 1\right) - V(|\phi|)\\ =& -2\beta^2\left(\sqrt{1+\frac{2}{\beta^2}\left(\frac{(\eta f'(r))^2}{4} - \frac{1}{2}\left(\frac{\eta a f}{r}\right)^2\right)}\sqrt{1+\frac{2}{\beta^2}\left(\frac{1}{2}\left(\frac{a'(r)}{er}\right)^2 + \frac{1}{4}\left(\frac{1-a^2}{er^2}\right)^2\right)} - 1\right)\\ & - V(f).\end{aligned} \tag{66}$$

Setting $e = 1 = \eta$ and substracting with

$$\mathcal{L}_{\text{BPS}} = -\frac{F'(f)A}{r^2}f' - \frac{FA'(a)}{r^2}a' \tag{67}$$

we have

$$\begin{aligned}\mathcal{L}_{\text{BPS}} - \mathcal{L} =& 2\beta^2\left(\sqrt{1+\frac{1}{\beta^2}\left(\frac{(f')^2}{4}-\frac{1}{2}\left(\frac{af}{r}\right)^2\right)}\sqrt{1+\frac{1}{\beta^2}\left(\frac{1}{2}\left(\frac{a'}{r}\right)^2+\frac{1}{4}\left(\frac{1-a^2}{r^2}\right)^2\right)}-1\right)\\ &+V(f)-\frac{F'A}{r^2}f'-\frac{FA'}{r^2}a'=0.\end{aligned} \tag{68}$$

Now naming $f' \equiv x$, $a' \equiv y$, $\text{Qf} \equiv F'A$, $\text{Qa} \equiv FA'$ so that we have

$$a'_{\pm} = \frac{\pm\sqrt{D_0} + \text{QaQf}r^2x - \text{Qa}r^4\left(V - 2\beta^2\right)}{r^2\left(4a^2f^2 - \text{Qa}^2 + 2r^2\left(2\beta^2 + x^2\right)\right)} \tag{69}$$

with

$$\begin{aligned}D_0 =& r^2\left(2a^2f^2 + r^2\left(2\beta^2 + x^2\right)\right)\\ &\times\left\{-4a^6f^2 + a^4\left(8f^2 + \text{Qa}^2 - 2r^2\left(2\beta^2 + x^2\right)\right)\right.\\ &-2a^2\left(f^2\left(4\beta^2r^4 + 2\right) + \text{Qa}^2 - 2r^2\left(2\beta^2 + x^2\right)\right)\\ &+2\beta^2r^4\left(\text{Qa}^2 + 4\text{Qf}x - 2r^2\left(2V + x^2\right)\right) + \text{Qa}^2\\ &\left.+2r^2\left(-2\beta^2 + \left(\text{Qf}^2 - 1\right)x^2 - 2\text{Qf}r^2Vx + r^4V^2\right)\right\}\end{aligned} \tag{70}$$

that must be zero. This implies that, since we want to find F and A, we need the terms inside the curly bracket to be zero. Then it produces

$$f'_{\pm} = \frac{\pm\sqrt{2}\sqrt{D_1} - 2\text{Qf}r^4\left(V - 2\beta^2\right)}{2r^2\left(a^4 - 2a^2 - \text{Qf}^2 + 2\beta^2r^4 + 1\right)} \tag{71}$$

with

$$\begin{aligned}D_1 =& r^2\left(\left(a^2-1\right)^2 + 2\beta^2r^4\right)\\ &\times\left\{-4a^6f^2 + a^4\left(8f^2 + \text{Qa}^2 - 4\beta^2r^2\right) + a^2\left(4f^2\left(\text{Qf}^2 - 2\beta^2r^4 - 1\right) - 2\text{Qa}^2 + 8\beta^2r^2\right)\right.\\ &\left.+\text{Qa}^2\left(-\text{Qf}^2 + 2\beta^2r^4 + 1\right) + 4\beta^2r^2\left(\text{Qf}^2 - 2r^4V - 1\right) + 2r^6V^2\right\}\end{aligned} \tag{72}$$

that must be zero. Again, since we want to find F and A, we need the terms inside the curly bracket to be zero. Since $D_1 = 0$ then the term inside curly bracket must be zero. Then expressing it as powers of r we have

$$\begin{aligned}&-\left(\left(a^2-1\right)^2 - \text{Qf}^2\right)\left(4a^2f^2 - \text{Qa}^2\right) + 2\beta^2r^4\left(\text{Qa}^2 - 4a^2f^2\right)\\ &\quad+4\beta^2r^2\left(\text{Qf}^2 - \left(a^2-1\right)^2\right) + 2r^6V\left(V - 4\beta^2\right) = 0.\end{aligned} \tag{73}$$

Since the BPS energy function is valid at all r, each term must be zero, which imply $V = 0$, $\text{Qa} = \pm 2af$, $\text{Qf} = \pm(a^2 - 1)$. Hence we obtain the standard BPS energy function $Q = 4\pi F(f)A(a)$ with $F = f$ and $A = \pm(a^2 - 1)$, and its standard BPS equations

$$f' = \frac{\text{Qf}}{r^2} = \frac{\pm(a^2-1)}{r^2}, \tag{74}$$

$$a' = \frac{\text{Qa}}{2} = \pm af. \tag{75}$$

7. Discussions

The method that had been proposed in [12] in obtaining BPS equations of some models of vortices is still able to obtain BPS equations of some nonabelian models of magnetic monopole. We are, at the moment, investigating the BPS equations in several other models using this method. Some of them (not shown here) refuse to give the BPS equations using our method, and thus we need to modify our formalism in order to enable it. For our next steps, we intend to find some other possible models that can be of interest, e.g., V dependent on $|\phi|$ from some other models since all above models have V = constant.

Acknowledgments

This research is supported by Research Center for Physics, Indonesian Institute of Sciences' Research Assistant contract No. B-453/IPT.1/HK/III/2017 and is partially funded by Universitas Indonesia's Research Grant from Internationally Indexed Publication for Final Assignment of Students of Universitas Indonesia (PITTA) No. 656/UN2.R3.1/HKP.05.00/2017.

References

[1] Kibble T W B 1976 *J. Phys.* **A9** 1387–1398
[2] Shellard E P S and Vilenkin A 1994 *Cosmic strings and other topological defects* (Cambridge: Cambridge University Press) p 64
[3] Rajaraman R 1982 *Solitons and Instantons: an Introduction to Solitons and Instantons* (Amsterdam: North-Holland) p 62
[4] Weinberg E J 2012 *Classical solutions in quantum field theory: Solitons and Instantons in High Energy Physics* (Cambridge: Cambridge University Press) p 89
[5] Dirac P A M 1931 *Proc. Roy. Soc. Lond.* **A133** 60–72
[6] 't Hooft G 1974 *Nucl. Phys.* **B79** 276–284
[7] Polyakov A M 1974 *JETP Lett.* **20** 194–195 [Pisma Zh. Eksp. Teor. Fiz.20,430(1974)]
[8] Polyakov A M 1975 *Zh. Eksp. Teor. Fiz.* **68** 1975
[9] Prasad M K and Sommerfield C M 1975 *Phys. Rev. Lett.* **35** 760–762
[10] Bogomolny E B 1976 *Sov. J. Nucl. Phys.* **24** 449 [Yad. Fiz.24,861(1976)]
[11] Atmaja A N and Ramadhan H S 2014 *Phys. Rev.* **D90** 105009 (*Preprint* 1406.6180)
[12] Atmaja A N 2017 *Phys. Lett.* **B768** 351–358 (*Preprint* 1511.01620)
[13] Nakamula A and Shiraishi K 1991 *Hadronic J.* **14** 369–375 (*Preprint* 1309.0282)

IOP Conf. Series: Journal of Physics: Conf. Series **988** (2018) 012007 doi:10.1088/1742-6596/988/1/012007

Modified Friedmann equations in braneworld scenario

Kevin F S Pardede[1], Agus Suroso[1,2] and Freddy P Zen[1,2]
1Theoretical Physics Division, Institut Teknologi Bandung, Jl. Ganesha 10, Bandung 40132, Indonesia
2Indonesia Center for Theoretical and Mathematical Physics (ICTMP), Jl. Ganesha 10, Bandung 40312, Indonesia

E-mail: kevinfrankly@rocketmail.com

Abstract. Firstly we present a brief discussion about braneworld model and the derivation of Friedmann equations in the braneworld scenario. Next we carry out a dynamical analysis of the modified Friedmann equations. Lastly, we perform a numerical computation using the Metropolis-Hastings algorithm to compute several free parameters of the modified Friedmann equations. Among others, we found that the brane tension is not large enough to support a nucleosynthesis process which correctly reproduce the abundance of present light elements. Also, we found a negative dark-radiation energy density, and thus indicating a possibility of non-singular initial universe.

1. Introduction

Modification of general relativity, has long been one of the main theme of research in theoretical physics. Although the success of general relativity of passing several test for its correctness is indisputable (for example: general relativity predicts the existence of gravitational waves that has only been recently detected), there are still lot of questions remain unanswered. One of those problems is that when we try to apply general relativity to cosmology, it is found that $\sim 95\%$ of our universe is unknown (the dark energy + dark matter). Our past experience regarding the need of inventing of somewhat artifical object (in this case the dark energy/matter) might be a clue that our theory needs to be modified (for example: the existence of aether isn't necessary after we find a better theory than Newtonian mechanics, which is the theory of special relativity by Einstein).

One of the branch of modified general relativity is to consider a conventional spacetime (3+1 dimension) with extra dimension (which usually is a spatial dimension). In the context of particle physics, the development of braneworld scenario (a model of universe, where a brane which is a hypersurface embedded in the full spacetime/bulk), has its roots from the needs to explain the gauge hierarchy problem (the large discrepancy between the planck Mass M_P and electroweak mass scale M_{EW}: $M_P/M_{EW} \sim 10^{16}$). One of the earliest model is the large extra dimension developed by Nima, *et al* [1], which basically "solved" the problem, given there are ≥ 2 extra dimensions compactified within a radius in the millimeter-order. The unusually large extra dimension predicted by the previous model, leads to a generally better model by Randall-Sundrum [2],[3] for the case of two and single brane respectively in a five dimensional warped spacetime, where the radius of the extra dimension is $\sim M_P^{-1}$.

A more general framework to obtain the formulation of general relativity in braneworld scenario is developed by Shiromizu, *et al* [4]. Assuming that the Einstein equation holds in

IOP Conf. Series: Journal of Physics: Conf. Series **988** (2018) 012007 doi:10.1088/1742-6596/988/1/012007

five dimension, the projection of the Einstein equation in the usual four dimension giving us the effective Einstein equation. The effective Einstein equation obtained differs from the usual Einstein equation in the sense that it is not closed (contain a terms that depend on the the bulk spacetime) and that it contains a term proportional to square of energy momentum tensor.

One way out this problem (non-closed equation) is a method called gradient expansion method [5],[6] developed by Kanno and Soda. This method assume that the energy density of the matter is small relative to the energy scale of the brane (in this case is called brane tension). The basic idea then is to solve the Einstein equation perturbatively with the ratio of energy scale of the matter and brane as the perturbation parameter [7]. Using this method, several generalization of Randall Sundrum braneworld model has been developed, for example the 3 brane model [8],[9] or the braneworld of brane with dimension (4+n) living in a (5+n) dimensional bulk [10],[11].

In this paper, unless otherwise stated, we use the following index convention: $M, N = 0, 1, 2, 3, ...n$, $\mu, \nu = 0, 1, 2, 3$; and $a, b = 4, 5, 6, .., n$, for n-dimensional spacetime. To further distinguish the four or five dimensional tensor, we use the following convention: ${}^{(m)}A^{\mu_1\mu_2...}{}_{\nu_1\nu_2...}$ = tensor A in m dimension spacetime. Also as usual, the subscript 0 means that we considered the present epoch value of the concerned cosmological object. For example H_0 means the present epoch Hubble constant.

2. Modified Friedmann Equations

Let $\mathcal{M}$ be our five dimensional spacetime whose coordinate is $x^M = (x^\mu, y \equiv x^4)$. Consider a five dimensional braneworld scenario composed of single spacelike brane located at $y = 0$. In the Gaussian normal coordinate, the metric reads

$$ds^2 = {}^{(5)}g_{MN}dx^M dx^N = dy^2 + {}^{(4)}g_{\mu\nu}dx^\mu dx^\nu. \tag{1}$$

The five and four dimensional metric: ${}^{(5)}g_{MN}$, ${}^{(4)}g_{\mu\nu}$, respectively satisfying the following expression

$${}^{(5)}g_{MN} = {}^{(4)}g_{MN} + n_M n_N, \tag{2}$$

where n_M is the unit normal vector with respect to the brane, satisfying $n_M n^M = 1$ and ${}^{(4)}g_{MN}n^N = 0$.

Assume that the Einstein equation holds in five dimensions

$${}^{(5)}G_{MN} = \kappa_5^{25}T_{MN}, \tag{3}$$

Now let us decompose the five dimensional energy-momentum tensor as follows

$${}^{(5)}T_{MN} = -\Lambda_5 g_{MN} + {}^{(4)}T_{MN}\delta(y), \tag{4}$$

where Λ_5 is the bulk cosmological constant. The four dimensional energy momentum tensor ${}^{(4)}T_{MN}$ which is localized in the 3-*brane* ($y = 0$) can be further decomposed as follows

$${}^{(4)}T_{MN} = -\lambda q_{MN} + \tau_{MN}, \tag{5}$$

where λ is the brane tension and the matter energy momentum tensor τ_{MN} satisfy $\tau_{MN}n^N = 0$ (τ_{MN} has no normal component relative to the brane).

Using the Gauss-Codazzi equation the five dimensional Einstein equation can be projected into the four dimensional spacetime, and thus after imposing the $\mathbb{Z}_2$ symmetry on this spacetime (with respect to $y = 0$), resulting the renowned SMS (Shiromizu-Maeda-Sasaki) equation [4],[12]

$${}^{(4)}G_{\mu\nu} = -\Lambda_4 q_{\mu\nu} + 8\pi G_N \tau_{\mu\nu} + \kappa_5^4 \pi_{\mu\nu} - E_{\mu\nu}, \tag{6}$$

where

$$\Lambda_4 = \frac{1}{2}\kappa_5^2(\Lambda_5 + \frac{1}{6}\kappa_5^2\lambda^2), \tag{7}$$

$$G_N = \frac{\kappa_5^4\lambda}{48\pi}, \tag{8}$$

$$\pi_{\mu\nu} = -\frac{1}{4}\tau_{\mu\alpha}\tau_\nu{}^\alpha + \frac{1}{12}\tau\tau_{\mu\nu} + \frac{1}{8}q_{\mu\nu}\tau_{\alpha\beta}\tau^{\alpha\beta} - \frac{1}{24}q_{\mu\nu}\tau^2, \tag{9}$$

while $E_{\mu\nu}$ is the projected five dimensional Weyl tensor ${}^{(5)}C^A{}_{BRS}$

$$E_{\mu\nu} \equiv {}^{(5)}C^A{}_{BRS}n_A n^R q_\mu{}^B q_\nu{}^S. \tag{10}$$

Now as an application in cosmology, we modeled our four dimensional spacetime as a spatially homogeneous and isotropic space, which mathematically means a spacetime with the Robertson-Walker (RW) metric

$${}^{(4)}g_{\mu\nu} = \text{diag}(-1, a^2\gamma_{ij}), \tag{11}$$

where $a(t)$ is the scale factor, and γ_{ij} is the metric for three dimensional maximally symmetric space which is parametrized by $\kappa = -1, 0, 1$ (respectively: hyperbolic, flat, and spherical space) Explicitly for $(x_1, x_2, x_3) = (r, \theta, \phi)$

$$\begin{aligned} ds^2 &= \gamma_{ij}dx^i dx^j \\ &= \frac{dr^2}{1 - kr^2} + r^2 d\theta^2 + r^2\sin^2\theta d\phi^2, \end{aligned} \tag{12}$$

where $k \equiv \kappa/R^2$, and R is the radius of the space with the metric γ_{ij} mentioned above. The matter in the brane is modeled as a perfect fluid with the following energy momentum tensor

$$\tau^\mu{}_\nu = \text{diag}(-\rho, p, p, p). \tag{13}$$

From the Codazzi equation, our chosen five dimensional energy momentum tensor gives the usual energy momentum conservation law

$$D_\nu\tau_\mu{}^\nu = 0. \tag{14}$$

Assuming the simplest equation of state

$$p = w\rho, \tag{15}$$

the following relation between scale factor and the energy density could be derived

$$\rho \propto a^{-v}, \quad v \equiv 3w + 1, \tag{16}$$

Using the energy momentum conservation law (14) and assuming the RW metric, the following relation could be inferred [12]

$$E_{00} = 3\mathcal{E}_0/a^4, \tag{17}$$

where $3\mathcal{E}_0 = E_{00}(t = t_0)$ for $a(t_0) = 1$. Because $E_{00} \propto a^{-4}$ resembles a term from radiation, this term is usually called dark radiation, where the "dark" comes from minus sign in equation (18). From (17) and the energy momentum conservation, we arrived at the modified first and second Friedmann equations

Modified Friedmann equations I

$$\left(\frac{\dot{a}}{a}\right)^2 = \frac{8\pi G_N}{3}\rho - \frac{k}{a^2} + \frac{\Lambda_4}{3} + \underbrace{\left[\frac{\kappa_5^4}{36}\rho^2 - \frac{\mathcal{E}_0}{a^4}\right]}_{\text{correction terms}}. \tag{18}$$

Modified Friedmann equations II

$$\frac{\ddot{a}}{a} = -\frac{4\pi G_N}{3}(3p+\rho) + \frac{\Lambda_4}{3} + \underbrace{\left[-\rho^2\frac{\kappa_5^4}{18} - \rho p\frac{\kappa_5^4}{12} + \frac{\mathcal{E}_0}{a^4}\right]}_{\text{correction terms}}, \tag{19}$$

where $\Lambda_4 \equiv \Lambda_5 + \frac{1}{6}\kappa_5^2\lambda^2$ is the effective four dimensional cosmological constant.

From the above equations, we see that the correction terms for the conventional Friedmann equations provided by the braneworld model come in two terms: the $\propto \rho^2$ term and the dark radiation term. One of the curious feature of the correction terms arises when we consider $\mathcal{E}_0 > 0$. In this case, assuming a matterless universe $\rho = 0$ and a vanishing cosmological constant Λ_4, the first modified Friedmann equations (18) yields

$$\frac{H^2}{H_0^2} = \frac{1+\Omega_E}{a^2} - \frac{\Omega_E}{a^4}, \tag{20}$$

where we have defined

$$\Omega_E = \frac{\mathcal{E}_0}{H_0^2} > 0. \tag{21}$$

Because $\frac{H^2}{H_0^2} \geq 0$, we have

$$a \geq \sqrt{\frac{\Omega_E}{1+\Omega_E}} > 0 \ \text{, for } \Omega_E > 0, \tag{22}$$

and thus indicating that the universe might has no initial singularity. Although the above arguments is oversimplified, it can be shown that even in a universe that contains matter, almost all cases of w (parameter for equation of state) leads to a non-singular initial universe (see e.g. [12]).

3. Dynamical Analysis and Benchmark Model

Consider the modified Friedmann equations I (18) for the single matter case

$$1 = \Omega + \tilde{\Omega} + \Omega_k + \Omega_{\mathcal{E}} + \Omega_\Lambda, \tag{23}$$

where

$$\begin{aligned} &\Omega = \frac{8\pi G_N}{3H^2}\rho, \quad \tilde{\Omega} = \frac{8\pi G_N}{3H^2}\frac{\rho^2}{2\lambda}, \quad \Omega_k = -\frac{k}{a^2H^2}, \\ &\Omega_{\mathcal{E}} = -\frac{\mathcal{E}_0}{a^4H^2}, \quad \Omega_\Lambda = \frac{\Lambda_4}{3H^2}. \end{aligned} \tag{24}$$

Next, from (19) and the following relation: $\frac{\dot{H}}{H^2} = -1 + \frac{\ddot{a}}{aH^2}$, we have

$$\frac{\dot{H}}{H^2} = -(1+q); \tag{25}$$

where the deceleration parameter q is defined as follows

$$q \equiv \frac{3w+1}{2}\Omega + (2+3w)\tilde{\Omega} + \Omega_{\mathcal{E}} - \Omega_{\Lambda}. \tag{26}$$

A thorough dynamical analysis for the parameters in (24) has been carried out by Campos and Sopuerta in the case of $\Omega_{\mathcal{E}} = 0$ [13] and for more general case $\Omega_{\mathcal{E}} \neq 0$ [14]. In this section, we only want to analyze the effect of the parameters unique to the braneworld model ($\tilde{\Omega}$ and $\Omega_{\mathcal{E}}$). Now, assuming that $\Omega_\Lambda = 0$ and $\Omega_k = 0$ (flat universe), we secure a set of autonomous differential equations for the modified Friedmann equations

$$\Omega' = (2(1+q) - 3(1+w))\Omega, \tag{27}$$

$$\Omega'_{\mathcal{E}} = 2\Omega_{\mathcal{E}}(q-1), \tag{28}$$

where $q = -\frac{3}{2}\Omega(w+1) - (3w+1)\Omega_{\mathcal{E}} + (2+3w)$, $1 = \Omega + \tilde{\Omega} + \Omega_{\mathcal{E}}$, and $A' \equiv \dot{A}/H$ for arbitrary A.

A fixed point is a point which satisfy $(\Omega', \Omega'_{\mathcal{E}}) = (0,0)$. We found the complete set of fixed points as follows: $(\tilde{\Omega}, \Omega, \Omega_{\mathcal{E}}) = (1,0,0); (0,1,0); (0,0,1)$. The complete classification of those fixed points is given in table 1.

Table 1. Fixed points $(\tilde{\Omega}, \Omega, \Omega_{\mathcal{E}})$ classification

	$(\tilde{\Omega}, \Omega, \Omega_{\mathcal{E}}) = (1,0,0)$	$(\tilde{\Omega}, \Omega, \Omega_{\mathcal{E}}) = (0,1,0)$	$(\tilde{\Omega}, \Omega, \Omega_{\mathcal{E}}) = (0,0,1)$
A	$w < -1$	$-1 < w < 1/3$	$w > 1/3$
B	$w > -1/3$	-	$w < -1/3$
C	$-1 < w < -1/3$	$w < -1,\ w > 1/3$	$-1/3 < w < 1/3$
D	-	-	-
E	-	-	-

note: A, B, C, D, E is stable, unstable, saddle, stable spiral, unstable spiral points, respectively.

As for the phase space solution near the fixed points is given in figure 1, 2, 3

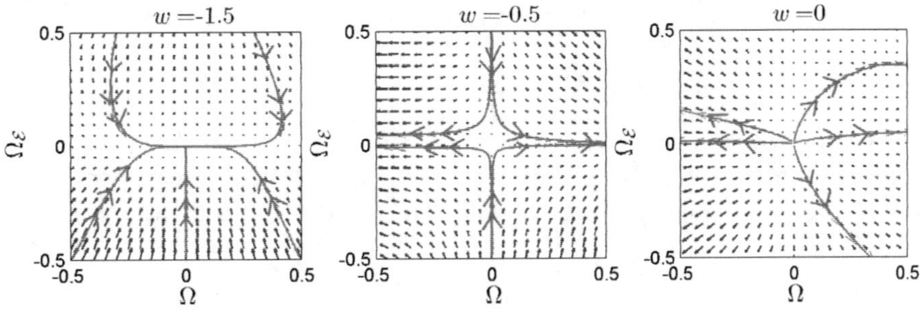

Figure 1. Phase space solution $(\Omega, \Omega_{\mathcal{E}})$ near $(0,0)$ for various value of w.

Next, we would like to compare (to fit) the modified Friedmann equations (18) with the Supernova Ia experimental data which has been compiled by Davis, et al. (2007) [15],[16],[17]. From the fitting process, we could determine the present epoch cosmological parameters value.

IOP Conf. Series: Journal of Physics: Conf. Series **988** (2018) 012007 doi:10.1088/1742-6596/988/1/012007

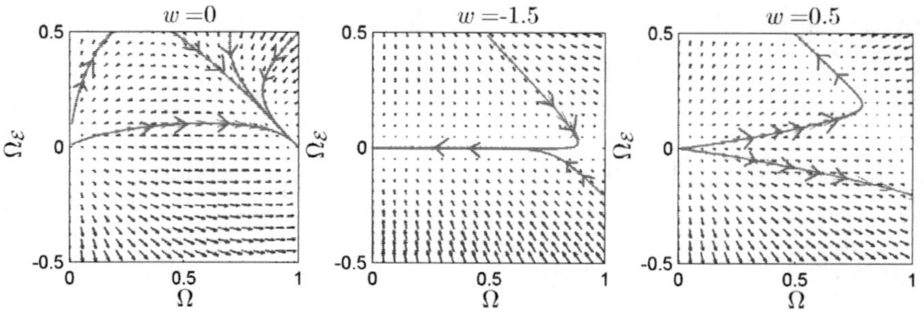

Figure 2. Phase space solution $(\Omega, \Omega_{\mathcal{E}})$ near $(1, 0)$ for various value of w.

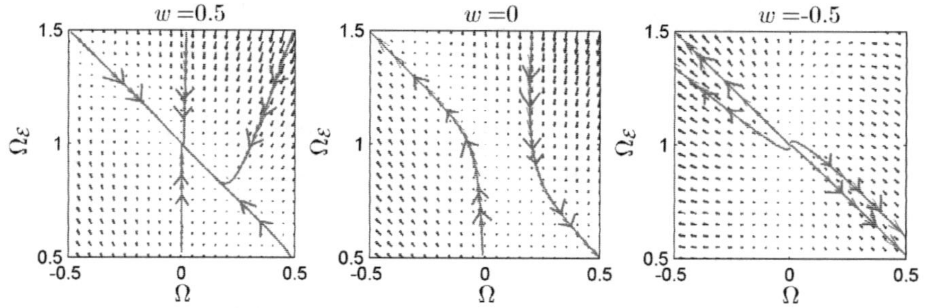

Figure 3. Phase space solution $(\Omega, \Omega_{\mathcal{E}})$ near $(0, 1)$ for various value of w.

Now, let us consider the modified Friedmann equations (18) composed of three types of matter: dust, radiation, and cosmological constant Λ_4. Thus, we have

$$\rho = \rho_{0,r}a^{-4} + \rho_{0,m}a^{-3}, \tag{29}$$

where the four dimensional cosmological constant energy density $\rho_{0,\Lambda}$ originated from the effective cosmological constant Λ_4, $\rho_{0,\Lambda} = \Lambda_4/8\pi G_N$.

Let $\mathcal{G} \equiv \frac{8\pi\lambda G_N}{3c^2}$, and define the following cosmological parameters

$$\Omega_{0,i} = \frac{8\pi G_N \rho_{0,i}}{3c^2 H_0^2}; \quad i = r, m, \tag{30}$$

$$\Omega_{0,\mathcal{E}} = -\frac{\mathcal{E}_0}{c^2 H_0^2}, \tag{31}$$

where subscript r, m stands for radiation ($w = 1/3$) and dust ($w = 0$) respectively. Assuming that the spatial curvature of our universe is zero, we could calculate the distance modulus as follows

$$\mu = 5 \log \frac{d_L}{10\text{pc}}, \tag{32}$$

where the luminosity distance d_L is calculated from

$$d_L = (1+z)c \int_0^z \frac{dz'}{H(z')}, \tag{33}$$

with the following expression for Hubble constant

$$\begin{aligned} H(z) = H_0\Bigg\{ &\Omega_{0,\Lambda} + \Omega_{0,m}(1+z)^3 + (\Omega_{0,r} + \Omega_{0,\mathcal{E}})(1+z)^4 \\ &+ \frac{H_0^2}{2\mathcal{G}}\left[\Omega_{0,m}(1+z)^3 + \Omega_{0,r}(1+z)^4\right]^2\Bigg\}^{1/2}. \end{aligned} \tag{34}$$

For simplicity, let us assumed that the contribution of total matter from radiation is neglibly small ($\Omega_{0,r} \simeq 0$), therefore we conclude that the parameters that we would like to find is the following: $\Omega_{0,m}, \Omega_{0,\Lambda}, \Omega_{0,\mathcal{E}}, H_0$, and the brane tension λ.

To find the best numerical value of those following parameters, we would like to minimize the following function [18]

$$\chi^2(\Omega_{0,m}, \Omega_{0,\Lambda}, \Omega_{0,\mathcal{E}}, H_0, \lambda) = \sum_{i=1}^{N_{\text{exp}}} \frac{[\mu(z_i) - \mu_{\text{exp}}(z_i)]^2}{\sigma_i^2}, \tag{35}$$

where $\mu_{\text{exp}}(z_i)$ is the experiment distance modulus for a data with redshift factor $z = z_i$ and error σ_i. Our compiled data, has a total number of experimental data $N_{\text{exp}} = 192$ [15]. The best fitted parameter has a characteristic of having [18] $\chi^2/\nu \simeq 1$, where ν is the difference between of total number of data and the degree of freedoms (number of parameters which we want to determine, which in our case is five).

In this work, we use the Metropolis-Hastings algorithm. Similar cosmological parameter estimation using this algorithm has been performed by Deffayet, *et al* in the case of DGP (Dvali-Gabadadze-Porrati) brane-induced braneworld model [19]. We found (this methods has been tested on the conventional Friedmann equations, which result can be found in the appendix)

$$\begin{aligned} \Omega_{0,m} &= 0.3453, \\ \Omega_{0,\Lambda} &= 0.6568, \\ \Omega_{0,\mathcal{E}} &= -0.0328, \\ \lambda &= 10^{35.5669} \text{ kg Mpc}^{-1} \text{ s}^{-2}, \\ H_0 &= 66.1527 \text{ km s}^{-1} \text{ Mpc }^{-1}, \\ \chi^2 &= 197.6811. \end{aligned} \tag{36}$$

Firstly, we notice that $\nu = 192-5 = 187$, and thus $\chi^2/\nu \simeq 1.06$. Next, we found that the present epoch Hubble constant H_0 is lower than the Hubble constant obtained from the conventional Friedmann equations [20] where $H_0 = 71.9$ km s^{-1} Mpc $^{-1}$. We also found that the numerical value of brane tension λ doesn't satisfy the lower bound required for a succesful nucleosynthesis phase [21]: $\lambda > 1\ (\text{MeV})^4 = 10^{49}$ kg Mpc^{-1} s^{-2}.

We also found that $\Omega_{0,\mathcal{E}} < 0$ which simply means that $\mathcal{E}_0 > 0$. It thus can be inferred from $\mathcal{E}_0 > 0$, that the evolution of our universe might has no initial singularity: $a(0) > 0$ [12].

Now because $\lambda \gg 1$ we can assume that $\tilde{\Omega} \simeq 0$, therefore we found that $q = -0.5169 < 0$. We thus conclude that the cosmology from braneworld scenario also predicts a universe which expands with positive acceleration.

Using the best fitted parameter we just obtained, the Hubble diagram for the modified Friedmann equation in the braneworld scenario can be compared with the experimental result, and also with the conventional model, as can be seen from Figure 4. As for the conventional cosmology, we use the following best fitted parameter obtained from the latest experiment: $H_0 = 71.9$ km s^{-1} Mpc^{-1} [20], $\Omega_m = 0.308$, $\Omega_r \simeq 0$, $\Omega_\Lambda = 0.692$ [22].

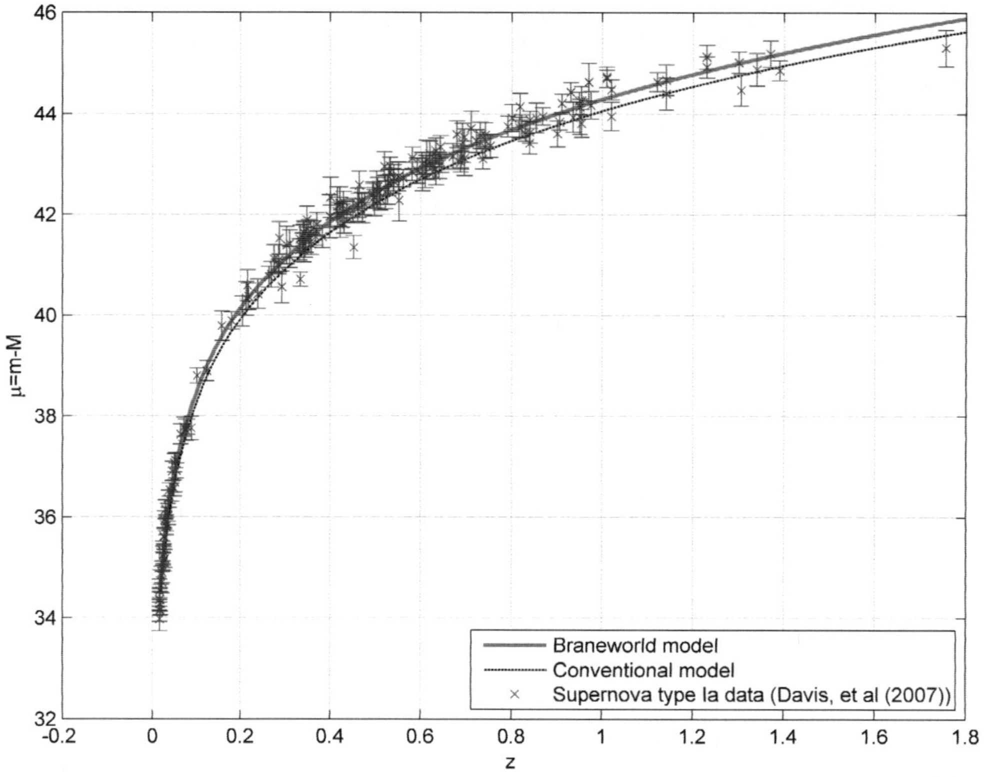

Figure 4. Comparison of Hubble diagram for braneworld model and conventional model with Supernova type Ia Data compiled by Davis, *et al.* (2007) [15],[16],[17].

4. Conclusion and Outlook

We found that the brane tension is not large enough to support a nucleosynthesis process which correctly reproduce the abundance of present light elements. Also, we found a negative dark-radiation energy density, and thus indicating a possibility of evolution without singularity. As for the outlook, it is either we further modify the Randall-Sundrum braneworld model or that we modify our numerical methods. One of the possibility of the modification of Randall-Sundrum braneworld models is to introduce a scalar field in the bulk [23] or in the case of universal extra dimension where the matter is moving freely in the bulk and does not confined in the brane: [24] and [25]. A more general theme of these modificaiton is to combine the Horndeski model (which it the most general, stable, scalar-tensor theory having second-order equations of motion) and the braneworld model. And for the simplest modification of the numerical methods we have used is to consider a better distribution proposal such as Gaussian distribution, or to use a certain convergence criterion.

Acknowledgments

The work of AS and FPZ is supported by Riset KK ITB 2017 and PUPT-DIKTI 2017.

IOP Conf. Series: Journal of Physics: Conf. Series **988** (2018) 012007 doi:10.1088/1742-6596/988/1/012007

Appendix

To find the best fitted parameters, we use the Metropolis-Hastings algorithm. In our case, the parameters space is five dimensional: $\mathbf{x} = (\Omega_{0,m}, \Omega_{0,\Lambda}, \Omega_{0,\mathcal{E}}, \lambda, H_0)$. Basically, we varied the parameters randomly to find the best fitted parameters with minimum χ^2. We varied parameters using a certain proposal distribution[26]. In our work, we use the simplest proposal distributions, which is the uniform distribution with a certain width $2\Delta x$ within the following interval $[-\Delta x, \Delta x]$. The proposed parameters will be accepted (the best fitted parameters is updated into the proposed parameters) if the proposed value of χ^2 is smaller, but also will be accepted with a certain probability in the case when the value of the proposed χ^2 is larger (as to avoid the local minimum of the χ^2 function)

In our work, the program will stop after a certain number of loops N. In principle, the results will independent of the initial guess, for a sufficiently large N. To obtain result (36), we set $N = 10000$ with the following initial value for the parameters

$$\begin{aligned} \Omega_{0,m} &= 0.3, \\ \Omega_{0,\Lambda} &= 0.7, \\ \Omega_{0,\mathcal{E}} &= 0, \\ \lambda &= 10^{50}, \\ H_0 &= 67 \text{ km s}^{-1} \text{ Mpc }^{-1}, \\ \Omega_{0,r} &= 0. \end{aligned} \tag{A.1}$$

Based on several initial trials, we set $\mathbf{\Delta x}$ as follows

$$\begin{aligned} \Delta\Omega_{0,m} &= 0.005, \\ \Delta\Omega_{0,\Lambda} &= 0.005, \\ \Delta\lambda &= 1, \\ \Delta\Omega_{0,\mathcal{E}} &= 0.01, \\ \Delta H_0 &= 0.01. \end{aligned} \tag{A.2}$$

As we have said in the main section, to ensure the validity of the result, this method has also been tested on the conventional Friedmann equation (where $\Omega_{0,r} = 0$ has been assumed)

$$H(z) = H_0 \left[\Omega_{0,\Lambda} + \Omega_{0,m}(1+z)^3 \right]^{1/2}. \tag{A.3}$$

For the conventional case we set $N = 1000$ with the same $\mathbf{\Delta x}$ as before (A.2). Assuming the spatially flat universe, and by using the following initial parameters value

$$\begin{aligned} \Omega_{0,m} &= 0.4, \\ \Omega_{0,\Lambda} &= 0.6, \\ H_0 &= 67 \text{ km s}^{-1} \text{ Mpc }^{-1}, \\ \Omega_{0,r} &= 0, \end{aligned} \tag{A.4}$$

we found the following parameters value for the conventional case

$$\begin{aligned} \Omega_{0,m} &= 0.2759, \\ \Omega_{0,\Lambda} &= 0.6880, \\ H_0 &= 67.1397 \text{ km s}^{-1} \text{ Mpc }^{-1}. \end{aligned} \tag{A.5}$$

References

[1] N. Arkani Hamed, S. Dimopoulos, and G. Dvali, "The hierarchy problem and new dimensions at a milimeter," *Physics Letters B* **429** no. 3, (1998) .

[2] L. Randall and R. Sundrum, "Large mass hierarchy from a small extra dimension," *Physical Review Letters* **83** no. 17, (1999) .

[3] L. Randall and R. Sundrum, "An alternative to compactification," *Physical Review Letters* **83** no. 23, (1999) .

[4] T. Shiromizu, K. Maeda, and M. Sasaki, "The einstein equations on the 3-brane world," *Physical Review D* **62** no. 2, (2000) .

[5] S. Kanno and J. Soda, "Brane world effective action at low energies and ads/cft correspondence," *Physical Review D* **66** no. 4, (2002) .

[6] S. Kanno and J. Soda, "Radion and holographic brane gravity," *Physical Review D* **66** no. 8, (2002) .

[7] F. P. Zen, B. Gunara, and H.Zainuddin, "The effective equation of motion on the brane world gravity," *Journal of Mathematical and Fundamental Sciences* **38** no. 1, (2006) .

[8] F. P. Zen, B.E.Gunara, and H.Zainuddin, "The low energy effective equations of motion for multibrane world gravity," arXiv:0511257.

[9] L. Cotta-Ramusino and D. Wands, "Low energy effective theory for a randall-sundrum scenario with a moving bulk brane," *Physical Review D* **75** no. 10, (2007) .

[10] S. Feranie and F. P. Zen, "Kaluza-klein two-brane-worlds cosmology at low energy," *Physical Review D* **81** no. 1, (2010) .

[11] Arianto, Freddy.P.Zen, S.Feranie, I.P.Widyatmika, and B.E.Gunara, "Kaluza-klein brane cosmology with a bulk scalar field," *Physical Review D* **84** no. 4, (2011) .

[12] K. Maeda, "The einstein equations on a brane world," *Progress of Theoretical Physics Supplement* **148** (2002) .

[13] A. Campos and C. F. Sopuerta, "Evolution of cosmological models in the brane-world scenario," *Physical Review D* **63** no. 10, (2001) .

[14] A. Campos and C. F. Sopuerta, "Bulk effects in the cosmological dynamics of brane-world scenarios," *Physical Review D* **64** no. 10, (2001) .

[15] T. M. Davis *et al.*, "Scrutinizing exotic cosmological models using essence supernova data combined with other cosmological probes," *The Astrophysical Journal* **666** no. 2, (2007) .

[16] Wood-Vasey, W.Michael, *et al.*, "Observational constraints on the nature of dark energy: first cosmological results from the essence supernova survey," *The Astrophysical Journal* **666** no. 2, (2007) .

[17] A. G. Riess *et al.*, "New hubble space telescope discoveries of type ia supernovae at $z \geq 1$: narrowing constraints on the early behavior of dark energy," *The Astrophysical Journal* **659** no. 1, (2007) .

[18] W. H. Press, S. A. Teukolsky, W. T. Vetterling, and B. P. Flannery, *Numerical recipes in C: the art of scientific computing*. Cambridge University Press, 1992.

[19] C. Deffayet, S. J. Landau, J. Raux, M. Zaldarriaga, and P. Astier, "Supernovae, cmb, and gravitational leakage into extra dimensions," *Physical Review D* **66** no. 2, (2002) 024019.

[20] B. V *et al.*, "H0licow v. new cosmograil time delays of he 0435-1223: H0 to 3.8% precision from strong lensing in a flat λ cdm model," *Montly Notices of the Royal Astronomical Society* **465** no. 4, (2016) .

[21] R. Maartens and K. Koyama, "Brane-world gravity," *Living Reviews in Relativity* **13** no. 1, (2010) .

[22] P. A. R. Ade *et al.*, "Planck 2015 results-xiii. cosmological parameters," *Astronomy & Astrophysics* **594** (2016) .

[23] M. Minamitsuji, "Braneworlds with field derivative coupling to the einstein tensor," *Physical Review D* **89** no. 6, (2014) .

[24] A. Suroso and F. P. Zen, "Varying gravitational constant in five dimensional universal extra dimension with nonminimal derivative coupling of scalar field," *Adv. Stud. Theor. Phys.* **6** no. 27, (2012) .

[25] A. Suroso and F. P. Zen, "Cosmological model with nonminimal derivative coupling of scalar fields in five dimensions," *General Relativity and Gravitation* **45** no. 4, (2013) .

[26] A. Heavens, "Statistical techniques in cosmology," arXiv:0906.0664.

Institute of Physics
Dirac House, Temple Back
Bristol BS1 6BE UK

ISSN: 1742-6588
ISBN 978-1-5108-6069-8